ADVENTURES IN AGRICULTURE

ADVENTURES

IN

AGRICULTURE

VOLUME 2

A COLLECTION OF
AGRICULTURAL PHILOSOPHY/HUMOR

BY

AUDRA BROWN

First Printing, 2017

ISBN 978-1-944256-09-8

Adventures in Agriculture
P.O. Box 5
Floyd, NM 88118
www.AinAg.audra-brown.com

Ordering Information:

Individual or quantity sales. Special discounts are available on quantity purchases of more than five copies. For details, contact the publisher via the info above.

Printed in the United States of America

First Edition

1   2   3   4   5   6   7   8   9   21   20   19   18   17

MORE FROM AUDRA BROWN

Adventures in Agriculture - Volume One
Tough Target (writing as Lee Brown)
429 to Yuma (Fire-for-Hire, Episode 1)
Hottest Mage Alive (Fire-for-Hire, Episode 2)

Check out her new columns online and in the
newspaper each week.

Go to:
www.audra-brown.com
for links, archives, enhanced content.

Also stay in touch by following her at:
www.facebook.com/ToughTarget

More info at: www.2epublishing.com

CONTENTS

"You can make a small fortune in farming-provided you start with a large one. "

- Anonymous

FOREWORD

Audra Brown may be the closest thing to a Renaissance woman I have ever met.

Writer. Musician. Farmer. Rancher. Black belt. Taekwondo world champion. Mechanic. Sharpshooter. Computer geek. Restorer of old cars. She was the kid who would save up money so she could buy---and I'm not kidding here---25 boxes of books from the used book sale at the local library to fuel a deep and vivid imagination through a winter in the country.

But asked to describe herself, Audra will say she is and always will be a farm kid. The adventures in this book all happened---or at least that's her story, and if I know her, she's sticking with it.

Audra is a gifted storyteller with an authentic voice seasoned with a splash of wit. When she tells you that "rain, oiled leather, and the smell of freshly turned earth have to rank as some of the best smells there are," you trust her. Whether she's bottle-feeding a newborn calf, busting ice on a tank, or heading to town for tractor parts, she's got room in her pickup for one more, and that would be our seat.

Pour that second cup of coffee into a to-go cup and come on.

Daylight's wasting.

> Betty Williamson
> Pep, New Mexico
> April 12, 2016

PREFACE

The profession of agriculture is both one of the oldest and one of the most advanced. Sometimes the fact that it's been around for thousands of years tricks us into thinking that it's more or less the same.

Sometimes it is. Sometimes it ain't.

There is one particular myth, perpetuated by popular culture in movies, TV, and other entertaining media, that these stories might argue with a bit. It's not nearly as dull and boring as you'd hope.

It's an adventure.

And I love sharing it with readers. I get to do it every week thanks to editor David Stevens, and to Betty Williamson, who recommended that he try me out as a columnist.

I am also indebted to my family. They are the reason I have these stories to tell and frequently the inspiration as well. They are both the causes and the victims of my success.

The biggest thanks, though, is to the readers. You are the best. Every time I get a letter, meet one of you in town, or just hear from a friend that you enjoy reading my columns, it makes my day.

Ya'll keep reading. I'll keep writing.

> Audra Brown
> Bethel, New Mexico
> April 13, 2016

To my Grandma Frieda,
an ag-adventurer of great skill and repute.

THE AGRIMYTHOS

Some stories have a lesson to tell. Some stories have a message to convey, and some stories just happen.

As we all know, even just after the fact, details and descriptions begin to blur, a phenomena that eventually results in the great myths that make some of the best yarns. This is one such muddied stream of events that somehow came to be...

Once upon a time, Mahindra, Kubota, and their little friend Fendt were chasing a deere...

They were young and ambitious and one day, while hunting for their elusive prey, they plowed headlong into what might become a legend. They chased their prey all across the Great Plains and even so far that they forded a river and crossed over into the land known as New Holland.

With this daring move, what had been just a local matter, was all of a sudden an international case.

The land was indeed great and wide and very versatile. They stumbled upon the lair of an old krone who claimed to be an oracle capable of seeing the future of the rains.

"What should we do?" the lost hunters inquired.

The witch cackled and said, "I see fahr and long, happy farmers. To be the ag-kings, first, you must defeat all the challengers, and last, you must farm all."

They waited for more, but the witch was done. She steigered towards her bedroll and yelled, "Schramm!"

They left with the east wind and an empire on their minds. They scouted the land and saw many creatures. There was a bobcat, a centaur, an eagle, several other kinds of cats, including one that looked kind of like a worm, but more than seemed possible, were the deere. They were many and seemed to be everywhere.

Some time passed, and then Mahindra laughed and looked at her companion. "Should we trust the words of such an ugly witch?"

Not so sure that they should ditch the witch, Kubota asked, "I mean, she could be mo-lean, but she mi-not be so bad. Don't you think there is some information to glean? What did you find so ugly, sister?"

All she had to say was:

"Oliver."

Audra Brown thinks she's pretty punny sometimes.

RURAL RELATIVE MEASUREMENTS

"I'm going out of town."

A common phrase, but if you hear a farmer or rancher say it, they likely mean exactly what you think they mean, but not what those particular words mean when taken literally. Whenever I need to tell someone I'll be significantly absent from my usual place, and I decide to use the standard phrase, a little internal amusement at the untruth of the words is inevitable.

"Out of town" is the definition of my home, not a rare or special place. But if I were to state that "I'm gonna be in town for a few days," confusion seems a likely result.

Better to lie and get my meaning across than to tell the truth that no one can understand.

Folks do like to measure and label places relative to town, don't they? Uptown, downtown, mid-town, China-town, outskirts of town, and the ever indefinable, bad-part-of-town.

I'm lost in a hurry.

As someone who will forever be from Out-of-Town, even when it comes to my hometown, I just nod like I know the difference and get a map.

There's a mysterious land that's known to outsiders as "the country" or sometimes "out in the county." It's where I'm from, and it's loosely defined as not-in-town and once you get far enough out, close-to-town isn't included either. It's a bit contrary, but we define places relative to town too. Probably because it's usually in the same place. Town is just easier to locate than country.

"I'm going to town" is a common phrase that is usually followed by an offer to pick stuff up for you so that you can put off your own trip to town. Places are in-town, close-to-town, out-of-town, the-other-side-of-town, west-of-town, south-of-town, east-of-town, north-of-town, and so forth.

General directions like that work if there's a big, obvious town close by that is clearly more notable than the competition. Not always the case, and a that's probably a reason why we tend to be really specific and actually name towns more often than not. Even towns that no longer exist.

My address is Portales, I'd claim Bethel as my hometown, and I'm part of the Floyd community. But I've never lived properly in any of those places.

I think maybe next time somebody asks, I'll just say I'm from Out of Town, see how that goes.

Audra Brown is going out of town after work today, just

FARMER FIDGET SPINNERS

This new fad that has swept the parking-lot vendors on both sides of the road; this thing they call fidget spinners...is interesting to me.

I mean, I can't argue that anything on a bearing (so long as it isn't froze up with rust, of course) is inevitably pretty entertaining. So, I'm both a bit surprised that it took this long to become obvious, and also a bit underwhelmed due to the fact that I think it's something everyone already should have known.

There's plenty of things to spin on the farm.

The gauge wheels on a plow---when it's picked up or put down such that they're treading only on air. If it folds up just right, you can sit in the middle and go around and around a few times.

While we're talking about tires, what's more fun that a couple of loose radials to roll around? You can race with a fellow boredom-avoider or challenge yourself and see how many you can keep going. If you have the time and the additional odds and ends, you can make ramps and rockers and all sorts of tricks for your tires to attempt.

Bearings alone are a good sort of pocket-sized fun, they spin and twiddle quite well---and if you grease them right, are silent too.

A ratcheting wrench is a heck of a thing to fiddle with, as you can alternate between twirling it like a mechanic's baton, and spinning it around your finger like those things that seem to have recently caught on. But, unlike the spinners that are for nothing but fidgeting, a racheting wrench can also be used to put together and take apart nearly anything.

The press wheel off a planter, the disk off a plow, the soil conditioner's cylindrical spinners, or the pulley off some old arm.

There are spinners to fidget with everywhere. But none of those are the official fiddle-focus that deserve the title.

The only thing that spins and distracts and keeps you occupied on the longest of days with nowhere to go and not enough to do...

...is the little knob that mounts on the steering wheel of any tractor worth driving, and would compete with any other device every invented for the number of idle-handed rotations.

The little thing that I grew up calling a "Spinner knob."

Audra Brown has twirled a "Brodie" knob recklessly and relentlessly.

FLAT EARTH ADVANTAGE

Regardless of your opinion on what shape the globe might be, there is incontrovertible evidence that there are some spots on this earth that deserve to be called flat. Places like the high plains.

I don't know how to explain all the things that make the flatlands the best home for this flatlander, but I'd wager that there's a little more than familiarity at play. It's true that I'm accustomed to and was raised where the earth is flat, but I've traveled and seen more than a few hills, trees, mountains and even an ocean or two.

I like the view form the top of a hill, love the endless waters of those oceans, but to be honest, when the trees and the mountains block my vision, I just get a little uncomfortable.

I've heard that the mountains and the trees can make a person feel safe and protected and that you can enjoy being out of the line of sight. But if I can hide, then so can everyone else. Something about it, I like to see where I am, where everybody else is, where we can all be going.

The funny thing is, when I can see where I might could be going, I don't feel near the hurry to get there.

But when there's a mountain in my way...

...I can't wait to see what's on the other side.

It makes me wonder if being a flatlander is about where you came from or if there's a certain sort of people that just like to see everything, regardless of the disadvantages or risks involved.

I'm gonna propose that a flatlander may not always be found on the plains. The earth must be more than flat, it has to be wide.

So, if you can't see the horizon, that doesn't mean you can't be a flatlander. My kind will be found on top of the hill, over the mountain, out at sea, or at the very least, up the tallest tree.

I don't know that there's a good reason or a bad reason to be a flatlander or some other sort, but I know that it's what I am and I'll always find the spot where I can see as much of this earth as possible.

And if the earth was flat...then there'd be some perfect spot where you could see it all.

Audra Brown will be seeing you.

TOYS OF CONVENIENCE

What is a toy but that object which brings you amusement via some exercise of your imagination?

Some toys are packaged and wrapped and labeled, intended for a specific sort of imagining and amusement. But others can be quite generic.

I prefer those with more amorphous potential.

The first thing that a young'un on the farm or the ranch learns is that you never know where you'll be---or for how long.

You might be in the pickup all day, or you might be on a horse. You might be in a tractor, or you might be at the barn. You might be at the feed store, or you might be in a hole in the ground.

So, the second thing you learn is that fun is where you find it and a toy is anything at all.

An abandoned sprinkler drop can be your magical staff of power, for the wizard you imagine being. The sand on the floor of the tractor is easily comparable to an etch-a-sketch. A two-by-four and a supply of sizable dirt-clods is great for a bit of batting practice. They are also more season-resistant than snowballs and can be thrown quite well once you learn the knack of it.

A rope thrown over a tree or a barn rafter is fun as long as it lasts. And don't ask me why but the bits that are removed by a steel-working punch are like heavy little Legos and my collection will never be big enough. Those things are great.

A piece of plastic pipe, in a length that seems suitable, can be your sword as you quest after a rabbit or some more imaginary goal.

An artistic inclination can be fulfilled by finding out where the chalk is stashed and decorating the walls, or any other metal surface that is available. And if you find the cattle-marking paint-sticks, the walls of the crowding-alley will never be the same. (Especially if you use the silver. It, as of yet, has not shown any signs of wearing away.

If you're a sci-fi fan and aspire to be a Jedi, you can build a lightsaber hilt from the bits of lost scrap that you can find on the ground outside the barn. It might be a collection of washers, bushings, and a few cool-looking plumbing parts, or it might be scraps of steel shaft and duct-tape.

Either way, it's not sold in stores.
It's handmade and collected.

Audra Brown still has her lightsaber hilt.

JUST LIKE RIDING A HORSE

I don't remember exactly the first time I was put on a horse, but it was before I started learning to drive, so let's say four years old.

The first challenge is getting onboard.

When the stirrup is about level with your face and you can't reach the saddle-horn from the ground, traditional mounting methods are not really an option.

Alternative one is the Environment Utilization Mount.

Find a fence, vehicle, rock, or other object that lets you start from a significantly elevated position relative to the horse. This works, but you have to have both the willingness to commit by jumping off or over (because horses never stand close enough to make it easy) and the faith that the target will still be there when you land. The other problem is that you better be ready to ride for a while and your hat better be on tight. You don't want to be out in the middle of a flatland pasture and have your hat blow off. Your main options are get off and walk you and your horse a likely long way back to a suitable environmental mounting position or you leave your hat to be lost (which is like losing your favorite toy, but worse) and get a sunburn.

Thus, we arrive at mounting method two.

The Cheater Stirrup.

As a vertically-underdeveloped cow-kid, this is the crowning addition to your tack. If you've never known the challenge, you may not be able to comprehend the torture it is to not be able to get on and off a horse at your leisure.

The cheater stirrup is probably an old stirrup that was laying around either without a mate or without prospects. It was unwanted and unloved until it got tied on the side of your saddle. It's probably still higher (relative to your body) than would make getting your foot up there easy, but it can now be done.

You grab the saddle leathers, get your foot in the cheater and suddenly you're on the side of your horse. A little foot switching and you've got your left foot in the real stirrup and your right leg swinging over to catch the other.

The only feeling that compares to sitting up on the back of a horse is being able to get there.

Audra Brown says that falling off a horse isn't nearly as bad if you can get right back on.

THE SCHOOL OF HARD NODS

How hard is it to refrain from scratching that itch on your nose? Now imagine that it's more than just an itch, it's a big, black fly that just landed on your forward cartilage extension. Your hand wants to just shoot up and get rid of the pesky pest.

But you can't, you're at an auction.

Delayed irritant removal is a critical skill that you learn early. You don't want to soothe your nose just to find out that you bid on a pallet of used tarp straps and ruined sprinkler tires. Don't make any sudden moves, turn away slowly, and in a similarly restrained motion, move your hand up to your nose and take care of business.

You learn a lot at farm auctions. I'd have to say that for a growing farm kid, they are one of the most concentrated educational environments. There is math to be done in your head. You have to quickly estimate the value of something. You take the potential value, modify that number with the probability of unseen problems weighted by the likely cost of those likely problems, should they exist, and then you've got a maximum bid estimate. Then you run that by your internal ledger of how much you've already bought at the auction, how much you actually have to spend, and how much you have left on your bank note.

Then you get into the quick stuff, recalculating with each bid, trying to decide how fast to bid, in what sized increments to bid, and how much over your original estimate you can really justify.

All the while, you're watching the auctioneer, the bid callers, and the people that are bidding against you. You've got to track the rhythm and content of the auctioneers roll, noting changes and predicting his next adjustment to increment or the final sold signal. You've got to keep your eyes on the eyes of the caller you're working with, keep his attention so that he immediately responds to your indications and gets your bids in. You track the other callers, judging their confidence and impression of the people they are catching bids from. And you use that information to add to the impression you're gathering of the opposing bidders themselves.

How much do they want it? Is that nonchalant confidence real or staged? What impression are they getting from you? Are they done? Or still in? Or just waiting?

And then it's over and your either won or not and you start the next calculations.

(And you can scratch your nose without consequence for at least the next few seconds.)

Audra Brown has bought some interesting stuff.

STANDING UP TO THE HERD

Stage fright, a phenomenon of cumulative social pressure and the expectations and judgments of a crowd of people, is an interesting concept.

It is considered a major barrier to many performing arts where the main venue is a stage in front of a live crowd.

I remember the first time I sang on a stage, the first time I broke a stack of bricks for a demonstration, the first time I got up the courage to play a guitar in front of an audience, and while there was a certain apprehensive excitement, none of those situations seemed to deserve the sort of fear that I'd heard about.

I've decided that it is a matter of perspective.

Orienting the situation as it compares to other experiences is a trick that I do all the time and it helps me push myself to do new things that, excitingly, are a little uncomfortable.

A perspective that I find useful in illuminating and overcoming the fear of standing in front of a socially-dangerous situation, is comparing it to the experience of standing in front of a more physically dangerous crowd.

...or perhaps a better term is herd.

Long before I ever put myself out in front of a bunch of people, who really only want to be entertained, I had to learn to put myself out in front of a bunch of cattle who really only wanted to not be there at all.

In both cases, an attitude of self-assurance and control makes everything work. But in one situation, the immediately likely consequence of being unconvincing is that they don't cheer.

In the other situation, they will run you over you if they don't find you suitably impressive.

I'm not gonna say that folks ought to have to be put out in front of a uncompassionate mob of animals that outweigh you, can outrun you, and outnumber you a hundred to one. But I am saying that a room full of fifteen skeptical humans just isn't the same level of intimidating if you've gotten comfortable working cows.

But if you need to get over that things they call stage fright, maybe this perspective will give you a little boost.

And if that doesn't work, go be a cowboy for a little while and see how scary people are then.

Audra Brown will stand up to just about anything.

NO LINES IN THE PARKING LOT

Parking lot etiquette is rarely spoken of other than to complain, but there does seem to be a few unspoken rules that most civilized people (unless they have an excuse) abide by when navigating the parking area. W

hen lines are clearly painted on the ground, the assumption is that one is expected to park the automobile somewhere between the markings and not let the tail-end hang out too far behind.

This is fine except for the fact that in many cases, one size does not fit all. If you drive a pickup of any size, or a car that is classic enough to still have manual steering, the more efficient parking lot lines are too narrow, too square with each other, and the rows are not nearly far enough apart.

I think that the best parking lots have a lot of different angles and approaches and so will work for most passenger vehicles that might arrive.

But there is a place that no standardization exists.

A place where lines would be ignored if someone were to attempt to organize the unstraightenable.

This is the sale barn parking lot.

The average vehicle is at least a large pickup and most of these rigs are hooked on to a stock trailer of some length from 16-feet to triple-axled, might-as-well-be-a-semi-load. You can't put them in even rows, even if you could control when and from what direction they arrived.

And they are not alone. Semis with full-length cattle-pots, a few cars, maybe a tractor (just because why not), and some days a converted school-bus with the back half made into a flatbed so it can pull a medium-sized stock trailer back to someone's ranch.

Yet despite all this chaos and lack of previously agreed upon rules and painted lines, the maneuvering is deft and accidents severe enough to be noticed are rare and unexpected. There's something about a trailer hooked on the back that will increase the driving-skill of most any farmer or rancher and decrease the sense of most other cars on the road...

But the best part of the sale barn parking lot isn't the creative chaos of the parking situation, nor the long conversations with old friends over the smell of manure.

No, the best thing I've ever found in the sale barn parking lot was tamales. Those were sure good.

Audra Brown learned to back up a trailer long before she learned how to park inside the lines.

MILESTONES

Graduations, birthdays, your first solo tractor-driving job...

Designated markers of time and accomplishment are important to us. They come in all sizes. Some are celebrated by many, some are celebrated by a few, and some are celebrated alone. Some are unique to the individual, but many are shared and understood by others, and a few are just on the farm.

Growing up in the field of agriculture, there are definitely some life milestones that are worth mentioning.

There are markers indicative of size and age. You'll never forget the first time you drove through a gate, the first time you drove by yourself to Grandma's house, the day when you could finally drive while looking over the steering wheel rather than through it, and the oddly uneventful day when you are finally authorized to drive into town to get parts and supplies all by your lonesome.

Strength is another characteristic that we like to measure. Rather than pounds, it is more apt to note measurements in the all-important universal unit of feed-sacks. (They weigh 50 pounds, if you need the conversion.) It is a special day when you step on the scales and find that you weigh as much as a feed-sack.

Though the more proud day, is a day sometime earlier, when you found that you were strong enough to lift a feed-sack and carry it to the trough. One is always striving to lift the next multiple of feed sacks that is greater than your own weight and more than the amount carried by your closest sibling in age.

In the late single-digits, when tractor-driving jobs are old hat, they are nonetheless limited until the achievement of certain feats of strength and common sense.

One moves up a level the day you are able to clutch a John Deere tractor. One advances again when you are strong enough (or smart enough) to lift and attach a tongue to the draw bar. One is in high-demand when you can roll the tarp on and off of a semi-truck trailer or the grain cart. (This can be accelerated if your folks take the time to extend the grain-cart tarp handle a few feet so as to make it reachable by those of shorter stature.

To be big enough (or deft enough) to climb atop a free-standing bale of hay, confident enough to jump across to another row of hay bales, tall enough not to need a box to reach the controls on the hydraulic cattle chute...

There is a long list, but that's all for today.

Audra Brown can reach all the controls.

TALKING DIRTY

The subjects of conversation are as varied as the combinations of people who might engage in such occasions of trading thoughts via our words. But despite this infinite potential variety, some topics are more likely and more common than others.

In my experience, and likely in yours, there are conversations to be found that seem inordinately long and involved for the topic at play. These un-understood discussions are so noted as to have become part of our standard set of stereotype features.

How long can ladies talk about shoes? Longer than me, barring the situation where the specific shoes in the spotlight (using the footwear label loosely) are jumping stilts that I'm determined to try out eventually.

How long can guys talk about the sport about which you care the least? How important are these things that they can take up hours of conversation that might have been dedicated to something important? Something like the latest iteration of string-theory, rather than the latest episode of nerd-ish comedy… How about the latest bill proposals in one congress or another? Maybe some words might be saved up to work on the problems that we as a nation, a species, or a planet could face?

Or even better, let's talk about dirt.

Have you seen that field he's breaking?

It's good country, red dirt with a good mix of clay and sand, and that plow is doing a job—I'm tellin' you. That shower we got put enough moisture down that it's sticking together about right and those furrows are rolling over just as pretty as you'd ever want to see. It's taking the ground good, pulling even, and after each pass there's not a weed in sight—just that fresh-turned earth, looking nearly as good as it smells.

I can't truly convey a conversation about 120 acres of dirt that has just been flipped over, because we just aren't here long enough to do it justice or dig that deep.

It's a good example of how meaning and importance is very dependent on context.

While I'll surely still find myself wondering why other people talk about the things that they do, I'll also remember that not everyone might find a big, flat, brown circle of dirt to be worthy of a discussion. They just don't know what they're missing---is all I can assume.

Audra Brown didn't even mention how straight the furrows were.

SPLITTING DROPS

Hallelujah, it's raining water. You might notice a slight uptick in the mouth-corners of your favorite farmer or rancher. There is only one thing guaranteed to have that effect and it's moisture falling from the sky. But don't fear, while they may be moving a little faster and more likely to tell a joke than a horror story, they will still want to talk about the weather. Rain or shine, wind or snow, dry or drowning, we always want to discuss how wet it might be.

Not in vague terms either. No, we want numbers, data! While the question might seem vague enough: "You got any of this rain?" Do not assume that the simple, obvious, seemingly sufficient answer of "yes" will be accepted. How much rain? A lot? A little? Still too nonspecific. We want to measure every drop. Here on the high desert, accuracy down to the tenth-of-an-inch is the industry standard. Don't make the mistake of rounding. Exact measurements only. An inch and nine tenths is never just "a couple inches" and even a tenth (known to the uninformed as barely-enough-to-measure) is worthy of notice and sharing.

And the instruments that capture and measure the blessed rain? They are also important. A rain-gauge is not as simple as a transparent vertical tube with a closed bottom, an open top, and some tenth-of-an-inch marks to make an external ruler unneeded.

There are reasonable four-inch tubes, optimistic 6-7 inch tubes, and delusional tubes that are even longer. There are tapered gauges and straight-sided gauges (and an eternal ongoing debate as to which is more accurate…) There are gauges with debris/bug guards and those who are open to the world. There are plastic gauges that will be cracked by the next rain and glass gauges that will be busted for no good reason or when you over-zealously smack them on the fence-post to get the dadgummed bugs and dirt out that is messing with your measurements and obviously soaking up the moisture and making it disappear.

Rain-gauging is a ubiquitous obsession here where it's high and dry. When accuracy is demanded, equipment is important.

Some parting thoughts on the matter:

Redundancy, redundancy, redundancy. One gauge is never enough because catastrophic gauge failure is only a matter of time.

Everyone's favorite free thing with a company logo is a rain-gauge and companies will be judged on the quality and style of the free rain-gauge that they choose to offer.

Audra Brown will take any metal-bracket/glass-tubed gauge you find. Keep the tapered plastic ones.

KINDS OF COWS AROUND

There are a lot of different kinds of cows. They come in all colors. Some have long legs, some have short legs, and some have oddly short legs.

Some sport a solid pattern, while others are a mix of colors, displayed as spots, patches, and speckles. Some have horns, some have better hind-quarters, some are of a delicate constitution, and some can live off dirt.

These features can be relevant when making decisions about herd management, but such features are only indicative of potential and what you can hope to expect. None of that matters if the cow doesn't do her job.

The primary behavior that separates good cows from bad cows is production. A cow works for you, and if that cow doesn't produce calves, it ain't working.

If the cow is just starting out, you might give it a second year to get its act together, but after that, much as you don't want to, you've just got to cut her loose. Not every cow is cut out to be a working bovine.

Then you've got the cows that make it past the first cut, and have been around for a few years. They produce and they know the pasture. It's real easy just to leave 'em and let 'em stay until they get too old or die.

I mean, why disrupt the status quo?

First off, you've got to get some new blood in the gene pool. If you don't stir things up now and again, your herd gets complacent. Too used to doing the same thing in the same place that they develop a weakness to change and when that hundred-year blizzard comes along or that new strain of pneumonia-causing bacteria shows up, they don't fare good.

Then there are the bad apples.

They do the job, but they can't be trusted. They'll run you up the corral fence, lead the herd astray, and to be honest, they'll raise untrustworthy calves too.

You leave 'em long enough, they start to rub off on the rest of the herd, then you've got a systemic problem and your only choice is to do a major culling and then go down and roll the dice on some replacements at the sale barn.

Yeah, sometimes the replacements are ugly, sometimes they don't work out, but it's better to gamble on some new cows that might make things better than keep the old one that you know will keep messing things up.

Audra Brown needs some new cows.

THE PILE OF JUNK

First Impressions can be misleading.

On occasion, that initial judgment call is right, sometimes it is plain wrong, and sometimes...it's both.

Let's play a game. Imagine you're out on the farm, bopping down a dirt road, and you come upon a towering pile of eclectic, but heavy pieces of stuff parked in the corner of a circle. What are you seeing? It looks a lot like a pile of junk.

And you're not wrong. That's what it is. The trick is the bottom piece. There is something disguised beneath a few random tractor-weights, used chunks of concrete (some still attached to steep-pipe posts), some spiky (but heavy) scrap metal odds and ends, and nice dressing of dirt and tumbleweeds that brings it all together.

Under all that weighty variety, there's just one of one other piece. There's a three-bottom, deep-digging, disk-type breaking-plow. A mighty piece of machinery on its own, the mass of miscellanea atop is sufficient to shadow the great plow. But the pile is not without purpose.

No, it's not hidden in order to hide it, it just needs the extra weight. You see, this plow is a monster, as mean as it is misshapen, and can perform feats that few other tilling devices could even dream about. With disks as tall as I was, it can cut into the ground deep enough to transform a field by pulling up dirt that hasn't been on top in a very long time.

It can really turn a field around, but it is no small feat to pull this beast, and more than a tiny task to turn it around. It takes two tractors. Two hoss tractors. The mail pull is a big, 8-tired, 4-wheel drive, monster tractor that dwarfs the little pile of junk that is the huge plow. Then, chained to the front of the biggest tractor on the place, is the biggest row-crop tractor you can find. The straight isn't too bad. As the leader, you just have to keep the chain tight, while riding one wheel in the giant ditch from the last pass, and psychically predicting when the rear rig will pull down and adjust your speed accordingly so as not to break the chain or get stuck yourself. No problem.

Then there are the corners.

You cross the last ditch and kick it up a few gears as you make a circle significantly bigger than the tractor in the rear, while keeping the chain both tight, at a good angle, and beating the other rig around so that you are headed down the next row and ready to slow down and pull as the back tractor comes out of the corner and the big disk gets put back in the ground.

Audra Brown's front tractor broke the chain with a steady pull, no jerking.

LOST AND FOUND

I've heard that finding a needle in a haystack is hard, but I've yet to come across a situation that produced the need to solve that particular conundrum.

It's not for lack of the items. Haystacks and needles are both common enough in the agricultural endeavors, I've found. Due to that lack of direct experience on the matter, I'll abstain from any definitive judgments about the puzzles produced by the juxtaposition of needles and haystacks. I have, however, come across more than a few puzzles that I can discuss with more authority, and will certainly commit to the opinion that there are plenty of enigmas that involve things that are hard to find on the ranch or the farm.

Finding things is sometimes a major endeavor. There are a lot of places where things might be. What might surprise you is the size of things that are lost. A needle is pretty hard to find in a haystack or any other decent-sized search area. A fifteen-foot, flat-bed trailer, on the other hand, would seem notably harder to misplace.

As an example of one of the more frustrating and inefficient tasks on the farm, locating a misplaced trailer is not something I would prefer to do again anytime soon. But, while rare, it is a likely recurring problem. Like any reasonable search, you start with a little deduction. Where did I use it last? What was the last thing we hauled? Whose pickup was pulling the thing? Where might one have unhooked it?

That last question narrows the area from tens of square miles to somewhere within, say, a thousand feet of any often traveled road. That's still a lot of ground, but far more doable. The other questions lead you to look in the places that they bring to mind before grudgingly moving forward with a less focused search pattern.

You drive more than a few miles, spend more than half of the day, and still that darn trailer eludes you. It's tiring work. The worst part, is that the size of the search area is not the most likely factor in not-finding your sizable target. No, the problem is that there are far too many places for it to hide.

Round two is spent retracing those possible locations, taking even more care to note piles of dirt, vegetation (one can dream), and other equipment that might overshadow the wayward trailer.

Round three is when you happen to head home and drive by the other side of the barn that you somehow haven't seen so far, and there it is, plain as day, now that you're looking at the right spot.

Audra Brown tries not to lose her trailers.

SIMPLE ENOUGH EQUIPMENT

Operating a large piece of equipment, or any size, really---is an exercise in precision and prediction. From a bicycle to a bulldozer, the operator's intent is always to maneuver in order to accomplish the goal as accurately as possible. But there are no buttons that just say "go here" or "do that." There are only controls that move some part of the machine, in some direction or another, and you have to figure out how any command/action is going to turn out before you can try it, because, you know, the bigger the equipment, the harder you can hit something by mistake.

A typical modern automobile, has three essential controls. Accelerate, Decelerate, and Turn the Wheels. If it's a stick, you have a couple more. Simple enough. It's a fairly intuitive control-scheme, once it becomes familiar. The kicker is the fourth variable that also needs to be predicted and figured into your plan---the one that you don't control. Other vehicles and their drivers make the whole thing a lot harder by adding some unknowns.

I'll admit that other vehicles are far less of a consideration on the farm. But we make up for it by adding controls and contraptions that keep us on our toes---and keep those toes dexterous.

Consider the ubiquitous tractor, the first equipment you learn to drive as a kid. The tractor's movement is controlled by a Steering Wheel, a Clutch Pedal, Left and

Right Brakes, a Throttle Lever, and a Gear Lever. Simple enough. But a tractor is defined by its implement and that take as whole other set of controls. If it's a 3-point attachment, you add up and down and you have to make sure you can predict and maneuver a tool-bar that sticks out straight a good distance to either side. Again, simple enough, because it can be worse.

Then you start hooking things on at the draw-bar, an articulating attachment point (sometimes connected to an articulating tongue). You now have to predict which way and how fast this attached vehicle will move depending on how the tractor goes. Then we get into the other levers that populate the tractor cab. Three or four Hydraulic Controls that can go two directions, and you have to remember which way you hooked up the hoses to know which direction on the lever matches which action on the implement behind you.

Then you have the PTO, the spinning shaft of great power and great doom. It has settings of On, Off, and Not Good!

Not enough? Some implements come with a box of new buttons that you mount in the cab. A sprayer adds a big box of buttons, the need to do frequent math, and a swinging, sixty-foot boom that no matter how good you are, will inevitably get tangled up in a fence.

Audra Brown hasn't hit anything with the sprayer in a while...

HUNTING BY COMMITTEE

As we've discussed before, varmints are perpetual pests that need to be controlled in number. Critters are less immediate problems, tending to be herbivores rather than possessing meat-tearing teeth.

But the critter can be a pest if its numbers grow too great. Critters both large and small have to eat, and while herbivores won't bite you, they will eat the grass that your livelihood-providing livestock also prefer to consume.

In order to prevent the critter population from exceeding the limits of the land, measures must be taken to keep the numbers down to a manageable level. With both varmints and critters, some methods of population management are more fun than others.

For large herbivorous critters, the parameters for removal are officially controlled. The method is that activity that is most commonly known as hunting.

There are rules and there are common practices. And then there are practices that are a bit odd.

Larger game hunting is known to be a somewhat individual endeavor. A silent, studious pursuit, either alone or maybe with one or two close friends. It is not often thought of as a spectator sport.

So, imagine, if you will, a hunt on the high, dry plains, where the hunter is out front, in a pickup or ATV with a driver/spotter/guide in the other seat. You head out in the morning, as soon as the sun rises and the mist clears, in search of a mature male antelope with horns that you hope are significantly taller than its ears.

It could be quiet and calm, but instead, you are followed by a procession of observers that are not subtle in any possible interpretation of the word. A deuce and a half with the back full of glass-wielding backseat-hunters who will spot you something to shoot from their elevated mobile perch can be handy for finding a shot.

But if you don't like the pressure of a crowd waiting for you to pull the trigger, you might experience some discomfort.

Hopefully, that won't throw your bullet off.

Hunting by committee is one way to do it.

And if you make that amazing shot, you've got plenty of witnesses to back you up.

On the other hand, you've also got plenty of witnesses if you .

Audra Brown did it and at least one mind was blown.

SNACKING OFF THE LAND

Ever get a case of the munchies?

Find yourself a few calories short in the long wait between regularly scheduled meals?

No problem, right?

Just bop down to the convenience store, the snack machine, or the nearest refrigerator.

But what if you're on the tractor in the middle of a farm, miles from the nearest fridge that still cools and contains anything other than a few parts that are better kept out of the weather. Bearings and their associated grease just aren't that filling.

In such cases, the very best plan is to not need one because you packed (and remembered to bring) a sufficient supply of snacks when you left the house this morning.

But alas, sometimes you are unprepared, forgetful, or your siblings came by and seriously decimated your snack stash. Then, the only options are to wait until someone brings you something (unlikely), go home (distant), or you resort to finding something edible in the nearby environment.

If you are strictly in a farming area, that can be difficult, but I guess in a pinch you can whip up a little tumbleweed salad if they are still in the green stage. Or, maybe you're working around a crop that is directly edible and you can pull up a handful---for quality assurance purposes, of course. The best bet is probably to check nearby vehicles for something left uneaten by yourself or another past driver.

This predicament also happens when you're out on the ranch. On a horse, it is admittedly more difficult to pack as many snacks. The best found munchibles, in my opinion, are sandhill plums, but those don't always make and they can be sneaky hard to find too. Next are probably prickly pears. Right time of year, you can peel the skin off those purple fruits and they are sweet, juicy treats. Pasture melons look like miniature versions of more common melons, but they really aren't that sweet. They're as good as how much salt you put on them.

When you're bored and hungry, you'll eat stuff that isn't all that appetizing, but it's all part of the experience, I suppose. So learn your field snacks.

There are some you can snack on and some that you can't. Know the difference first and then try the edible selection when you have no other choice.

Audra Brown has had some good snacks, and some bad snacks, and will probably try some more.

THE PICKUP WITH ALL THE DENTS

If your motor-vehicle of record never leaves the paved
road and only has encounters with those of its own kind
and the occasional stationary object that---of course-
--just leaps out in your way, then the stories it tells are
usually known as accidents and are hopefully forgotten
and fixed rather than displayed and told tales about for
years and years down the road.

The first thing about vehicles, is that contrary to
common belief, no large concentration of motors is
required to make dents, wrecks, or run-ins increase.

One vehicle, out alone on the farm, is perhaps as safe as
it will ever get.

With two, the inevitable result is that at some point they
will be in competition, and at some point they will hit
each other.

This takes a while, but I do believe it is as close to certain
as any physical law.

I speculate that the first two cars ever driven were first
admired, then raced, and then somehow met head-on.

I know at least one pickup that's certainly adorned with
the badges of adventure and a few clues that are related
to mysteries yet unsolved.

Dents from off-road vehicles, cracks in the grill from that time when the only two vehicles for miles slid through the sand and collided head-on, a slight twist to the rear-bumper from an adventure that involved being tied to a 2000-pound bull, and more scratches than you can count from grit, brush, and at least one angry, not-poled, yearling steer.

Another rig that comes to mind, has only one story to explain the many large dents. But any story that can beat up a pickup like that in a few minutes of a particular day, probably deserves all the credit that it gets.

Crazy 800-pound steers, are thankfully not the most common, but are common enough to guarantee at least one dent. This one was a killer, frothing and nuts. He rammed every door, making them notably thinner. He rammed it head on, making the hood a more difficult shape. He even did most of it over again just to make sure he really beat and scratched the red paint to make his point.

So if you ever see a pickup that's too dented to even make sense, it might have some stories that happened off the roads somewhere. At least when a car hits you, they only do it once and from only one direction---and cattle don't have good insurance.

Audra Brown's pickup has more stories than she'd like, but all the doors still open!

PICKING PEOPLE UP

Transportation is critical in any walk of life. To do much of anything, you need to get to where you are going.

Crawling is the method of choice for babies---and for when the house is on fire and you don't want to breathe so much smoke.

Walking is choice if the distance is short, the time is long, or there's no alternative.

Running is for emergencies and races only.

Unicycles are for clowns and people who like to learn things for no reason.

Bicycles are not for when you travel by sandy road.

Tricycles are for comparing with friends when you're not yet too tall to ride them.

Motorcycles are for going fast, driving over things that are not good for that purpose, and getting into wrecks while having way too much fun.

Tri-wheelers are four-wheelers for when you want to break your back.

Four-wheelers are for when the cattle are too ornery to mess with on two wheels.

Cars are for driving grandma back from church on Sunday, for looking cool, and thinking about going fast.

Pickups are for everything except metropolitan traffic.

Tractors are for driving in methodical patterns all day.

Semi-Trucks are for hauling on the highway, and probably not for moving hay out of sandy fields and down un-maintained roads...

Little planes are for spraying and moving yourself.

Big planes are for long trips and reminding yourself why you live in the desert where no one can find you that doesn't already know here you live.

Boats are for places with water, so I couldn't tell you what to do with them.

Ships are bigger boats and not my wheelhouse, again.

But despite all these and more that are capable of getting us where we need to go, you'd be surprised how often you end up somewhere, broke-down or otherwise in need, and wanting a ride.

And who do you call? Taxis don't run down the rutted track we call a road between the fields we so inventively call Number One and Number Three. As much time as a farmer spends alone running a tractor back and forth across the field, or a rancher spends out on his horse checking the fence, it's mighty important to have friends with a pickup and a key to the gate, who will come and get you when where you are going is not where you currently is.

Audra Brown's friends know her number.

EXTREMELY IMPROVISED SPORT

The life of a farm kid can be quite exciting, but it can also be interspersed with unbearable intervals of being stuck in a location with nothing in particular to do.

Innovation without preparation is a skill that we learn early in our careers.

What good is the sand that blows all over the fences and in your face?

It can be the best alternative to the snow we don't get to see when it has developed into a tall enough hill ---hopefully with minimal vegetation to impede your downhill path. The optimal angle depends on how fast you like to go and how much room there is to crash at the bottom. Beginners can ride those plastic discs down if they don't mind the possibility of a little spin. For the crazier kids, a snowboard can be much more directionally controlled, but also can be more likely to not make it all the way down as intended.

What is there to do when you are at the corrals and there is nothing but a manure pile and time to kill?

You find some way to slide down it. If you happen to have left your sandhill snowboard in the back of one of the pickups that you used to get there, you can do it in proper style.

When you weren't that conveniently supplied, a cowpie fight can be an exciting endeavor.

You'll find that the motivation to avoid getting hit is much more serious than in a simple dirt clod battle.

And the fun doesn't have to stop when you get back to working the cattle.

If you ended up with the job of pushing the calves up the alley into the shoot, you'll find yourself with time to kill and fun to devise. If the alley has suitable overhead pipes to grab and get yourself up over the calves, you can surf the bovines in lieu of having to use such an amateur method as poking them with a hot shot.

You can try and catch the goldfish that are swimming around in the tank, an amusement that will figure out if you have fast hands.

You can climb the sheds and equipment that happens to be nearby. You can paint murals on the crowding alley walls with paint markers that are supposed to be for marking a bovine head.

Amusement can be made of of anything, even a pile of what cows leave behind.

Audra Brown has also roasted quite a few fresh mountain oysters just to see how they'll pop.

TO TRIP A WALKER

In an effort to be overly prepared for the zombie apocalypse, I've been pondering the qualifications of various hay-bale-tying materials with regards to which would make the best post-apocalyptic accompaniment.

Undoubtedly, like many things overlooked and under-appreciated in the good times...twine, baling-wire, and their latest pal, net-wrap will be hoarded and treasured in a zombified world. So, let's be about it and see what perils these menacing materials might be capable of.

Baling-wire is classic, it is best known for wrapping itself around drive-lines, axles, and anything else that moves in a circular fashion. It's main enemies are side-cutters and rust. Of all its brethren, it stands out as the most commonly useful in a non-destructive way. It can be used to fix tools, tie gates, makes toys, and pull corks out of bottles---just to list a few.

Net-wrap is a newcomer, the controversial usurper that has taken the throne in the realm of round-bale wrapping. It is the embodiment of a philosophy of strength in numbers, containing large bales of hay with no knots, only two layers of thin, plastic, mesh. It would be great for trapping/tripping up hordes of bipedal enemies such as the walking dead. Even with the sharpest knife, it is not a quick or easy process to cut through a mess of net-wrap.

Finally, there is the bane of all existence...

...orange twine.

It has the tying power of wire and the tangling talent of net-wrap; able to shut down large machines with a single mound. It is long-lasting, plentiful, and is certain to cause pain and ball-ups wherever it is encountered.

I believe that it is destined to be a fearful weapon in the future war on the undead, but until that time comes, I think that lives are better spent in the vain attempt to eradicate it wherever it is found.

After all, is such a nuisance really worth the uncertain benefit of being as dastardly to theoretical monsters of the future as it is to the present day farmer having to set his treader-axles on fire to melt it away? Or the Cattleman who trips on the mounds that litter his corrals?

I leave it to you to decide. Personally, I will consider an alliance with wire and net-wrap long before I will be able to bury the hatchet with orange baling twine.

Audra Brown says, "Burn it! Burn it with fire!"

LIFE FROM THE FLOORBOARD OF A TRACTOR

The back of my new book makes the claim that I grew up in a tractor-seat instead of a car seat, and that's mostly true. In fact, I'd wager that my time spent in an actual car during my childhood was statistically equivalent to not-at-all. As a baby, I suspect you'd find me in pickups, tractors, combines, semis, parts stores, and Grandmas' houses.

A baby farmer spends many hours in the floorboard of Mom's tractor, oblivious to the world. Then they can walk and their domain grows to include the spacious and exciting (for a toddler) area behind the seat of a John Deere tractor. High over the world, and with the help of a couple of pillows and a blanket, the perfect place to take a nap.

Eventually, the small farmer takes a great step and move beyond Mom's tractor. Grandpa's combine, Daddy's truck, Uncle's airplane. There are many great days ahead in the not-quite-a-seat next to the driver of some big machine.

The aspiring farmer doesn't even realize how much is being learned, but will likely be able to tell you who has the best snacks, who will bring an extra coke, and where they hide it.

Soon enough, the new farmer gets the chance to be more independent and when left alone, finds out how much he/she actually knows. Puzzles and mysteries begin to have answers as the education of riding-along coalesces into applicable knowledge. Tricky machines with lots of buttons that only Daddy could fix gain a new trouble-fixer that can make it work, but may need a box to stand on to reach the right switch.

When problems out of the mini-farmer's reach come up and Dad needs to go take care of it, the combine doesn't have to stop running because the relief operator is sitting in the passenger seat, ready to go as soon as something heavy joins them in the Captain's chair to trick the sensor into thinking that they weigh enough.

True independence comes in the late single-digits when the pickup is mastered and driving faster than fifteen miles per hour lets you get around easier. No more just driving through the gates or inching the trailer along, you can go to Grandma's for supper and don't have to wait for someone to pick you up in the field when it's time to quit for the day.

But the scariest achievement, by far, is that teenage day when you drive home from church—in your grandparent's car.

Audra Brown didn't know a vehicle could be so low.

NATURAL TERRAIN

There's a terrain that runs through this part of the world and guess what, it's flat.

There just ain't much other way to describe it. It's a pretty plain plain. I'm personally quite fond of open country. Born a flatlander, I get a little claustrophobic when I can't see the curve of the earth by standing on top of the tractor.

The word terrain has some interesting connotations, from the simple relationship to its Greek ancestor to the idea that it is often used in the context of the tactical implications of an area. It almost implies a certain amount of vertical variation, in my experience.

We talk about rough terrain a lot more than we discuss open terrain. I looked it up and most types of terrain seem to be defined by features. Hills, ridges, mountains, streams, valleys, and so on and so forth. The typical terrain on the high plains is pretty much the absence of any of that.

It's always been interesting to me, how much the terrain, the topography, and the climate in a particular geographical area influences the behaviors of the people who live there. I figure it might be a good reason why the folks on the high plains tend to have an accent that's easy to holler with.

Getting the message across a wide space, likely with the wind blowing, is a needed skill. Certain sounds, rhythms, and emphases just work better in such a situation.

Alas, I was getting theoretical. I do that. How about something a little less nebulous? One of the advantages of level terrain is that you can see a long way. That means you can see things coming, but similarly, they can see you. There are times when you don't want something to notice you're coming. Like when you go hunting.

In other terrains, folks use the mysterious concept of natural cover to sneak up on the target. Here on the flatland, we just have to learn to shoot further. But, conversely, we're kinda straightforward and not very good at either the waiting or the sneaking that hunting in some other terrains requires.

Someone asked me the other day if there was any place around the range for some "natural terrain" shooting setups. They were implying a setup where there were natural obstacles, backstops, and otherwise interesting features to navigate and shoot in and around. The only answer is that there's plenty of natural terrain anywhere you look, it's just flat.

Audra Brown forgets sometimes that flat is not necessarily common.

MANUFACTURER RIVALRIES

What's better? Ford or Chevy or Dodge? Lamborghinis or Ferraris or something made by Carol Shelby?

Folks have opinions of varying depth and vehemence, and some are more right than others, but lemme tell you, if you want a real debate, just ask what color their tractor is and say that another color is better.

Most farmers I know have gone through a variety of brands of pickup over the years, usually having a preference, but willing to keep an open mind and make a decision based on the available features and which company hadn't stop offering a decent-sized gasoline engine that year. Heck, they might even make a deal based on which one they already had a flatbed to fit---or, if nothing else was left, the color choices might be important.

Tractors come in different colors too, but it's never an option.

The color of your tractor is a critical choice. John Deere Green, International Red, New Holland Blue, Cat Yellow, Steiger Green, or something less distinct if you're looking at one of those newer varieties that recently appeared or used to only be sold overseas. Unlike personal vehicles other than a DeLorean, the color is the brand, and the brand is the color.

No one confuses John Deere Green for anything else, and although you can get close with an off-brand can of almost matching paint, it's just not quite the same as the real hue. International Red goes well with the color of a Sunflower plow, but they are not quite the same, as close as they are. And construction equipment yellows? Don't mix 'em up. Caterpillar yellow is not to be confused with the color of Case—or the John Deere color that isn't green or a more bright, primary yellow.

Tractor colors don't mix, and while paint may fade with time, a farmer's loyalty to the color is a lot harder to change. A true John Deere man is identified by a shed full of nothing but green. A more open-minded farmer might mix it up across purposes. He might prefer International Harvesters, and New Holland swathers, and Bobcat for his skidsteer, but I'd bet a tank of diesel that he tries to keep his row-crop tractors the same shade of green.

You know, I've heard of culture shock and compromise when two folks get married. There are big decisions to make like what church to join together, how much money to share, and what names to keep. But there's nothing quite like growing up on different colored tractors and having to decide between a future that is red or green.

Audra Brown welcomes those who convert to the green side, but likes to keep her combines red.

THE BRAND OF A MAN'S HAT

The top of a person's head can tell you a lot about 'em and what they're up to that day.

A cowboy hat certainly stands out these days, but look closer. Felt hat or straw hat? If it's jacket-wearing weather, felt is fine, if it ain't, then something might be up. Is it a nice, stiff, clean hat? Then I'd wager that they're headed to or coming from a formal occasion. Is it a little soft, have a thick band of sweat-adhered dust around the middle? Does it have an awesome hatband that you'd wear everyday too? You might be looking at one of those folks cool enough to wear a felt hat year-round.

You can't tell much about their intentions. That hat on their head basically indicates that they are not an alien-impostor, but not much else.

Business as usual.

If you know that hat, but see it on the wrong head, then it's time to worry.

Felt cowboy hats can be any color as long as it's between white and black or some shade of brown. Somebody with a clean, white hat---really likes to wear a white hat. Some might say they're likely to be a better person.

I wouldn't know. I wear a black one.

Truth is, you probably shouldn't judge a person based on the color of their hat---the brand, however, is fair game.

Whatever the color, it's important. Hats are not like clothes. They're more like jewelry. Good jewelry. The stuff with diamonds and gold in it. The kind of stuff you maybe have one or two of and you keep it all your life and pass it down to your kids.

You don't talk about your hats in the plural. Each is unique and special and does not deserve to be lumped in with any other.

You don't buy a new one every year, you get one and you keep it a long time. If you replace it maybe once in your life, that old one probably gets passed on to be someone else's for a few decades. A cowboy hat isn't cheap, it's not easy to wear, it's important.

In fact, "hat-room" is a very important feature to consider when buying a new set of wheels. If you can't drive with your hat on, you won't be driving it.

Audra Brown is of the opinion that felt hats come in two types: Resistols and work-hats—black is implied.

RUNNING TOWARDS THE FIRE

There are a lot of requirements to being a farmer or a rancher. You have to have some real estate suitable (or at least potentially so) to your particular endeavor. You need equipment, time, a minimum level of cooperative weather, and the willingness to gamble year by year. You probably need some help and some idea of how to raise livestock or those plants that we call commodities.

Certain behaviors and characteristics are important as well. You'll definitely need a deep streak of stubborn and you better be (as they say) "a self-starter." Expect to improvise, spend lots of time alone with your own thoughts, working long days, and not maintaining a regular sleep schedule.

It's good to be judiciously cautious and think things through before you get too deep. But you need to act in a timely manner as well. Balancing immediacy and efficiency is an ongoing trial.

Except for when the fertilizer encounters the air-accelerating device. Then, you've got to drop everything else, grab a shovel and a wet bandana, and run as fast as you can towards the fire.

The fire can be a metaphor for any sort of trouble that you didn't see coming and every moment of delay is an increase in disaster.

When the cows are water-starved, dying, and out of
their minds---you've got to get them some water. When
the sprinkler is running away and getting ahead of itself,
you've got to get it shut down.

But time and again, one of the most frequent fires that
has to be put out is just that---a fire. You know...flames,
smoke, being hotter than a engine block in the summer.

There was one just last night, and run to it, we did.
Rolled in from town, saw the flames just up the road,
dancing bright and higher than a sprinkler tower.
Grabbed a wet bandana to throw over my nose and went
to grab a shovel.

Alas, the shovel took a few minutes to acquire, as the
one that is supposed to live in the back of my pickup
done R-U-N-O-F-T. But after only a minor delay, I had
four-wheels, a wet nose, and an entrenching tool.

Running to a fire is a race. A race that everybody loses
if no one makes it to the track, but luckily, most of the
time, it's a race that we just run against each other to see
who makes it before the faster folks have the fire put out.

*Audra Brown didn't win the race last night, but is glad
that someone did.*

COUNTRY COMPETITIVENESS

Farming and ranching is often full of surprises that keep us on our toes and busy, engaged, and entertained for a good deal of most of the time and then some.

There is a not insignificant amount of time and work that is not much more than what you might call routine. It does not always engage the mind and leaves us with time and motivation to expend on something else.

That something else, of course, has to be doable in amidst the day-to-day routine of fixing fences, driving tractors, checking water, and putting the cows back where they belong---among other things.

There is at least one good competition that meets not only these requirements, but has the added benefit of producing a useful result in and of itself.

Finding and removing rattlesnakes is what I have in mind. This is an old and storied tradition that manifests in different arenas and ways, but I'd bet more than a few rattles that at least one clearly defined competition of this sort is going on as we speak.

There are different rules that you can use to play the game.

You can keep track of the count, which is simple.

You can take note of the ingenuity that is used to take care of the situation and give bonus points for particularly creative solutions.

There can be a consideration for accuracy as well. If the head cannot be found after the first and only toss of lead in the snakes direction, that is worthy of at least a modest brag every now and again.

If you can knock it unconscious with just the passing movement of air and then smother it with a blue paper towel---then you'd better have skinned that unmarked hide and be able to show it to me and the other competitors. But if you did, then I guess you probably win the current season by the awesomeness clause.

I'll admit to not having a very high score so far this time around. It's a big, fat zero at the time of writing this. I'd appreciate any leads that anyone can share. I need all the help I can get because I'm already late to the game.

The kid is batting at least five the last time I heard, and the reigning champ is already at twice that.

You'd of thought he'd have run out by now. Or maybe he's shipping them in...

Audra Brown could pay more attention to the game, but then she might get a little too determined to win.

DOING IT THE EASY WAY

The basics of what needs doing haven't changed over the ages of agriculture. Farming requires that you grow plants and ranching requires that you raise livestock.

But thank goodness, even though the end-results are still pretty similar to the good old days, human ingenuity has produced new and less-exciting ways to do the job.

That is the trade-off that is hardest to feel good about sometimes. I mean, what would the stories about old cowboys be like if they had it this good?

Let us look at the compromises we modern cattlemen make in the interests of safety, reliability, and convenience. I mean, what else is there to do with all the extra time on our hands?

Laugh with me. That was a joke.

We're still busy and never get it all done, but, on a related thought, we probably, on the balance, have more hands ---and fingers.

I've spent my time on a horse, but a time came and went and now riding anything that doesn't rely on some refined petroleum combusting is more pleasure and diversion than actual work.

There are still places that a horse can go that a rig with at least two wheels cannot (and a unicycle-cowboy is still in the conceptual stage, thank the good Lord above.) Obsolete isn't the word yet, but rarely-needed is apt.

Once the cattle are gathered, probably while you listen to the radio, you again have a choice of tools and methods to get the 'working' done. Working cattle is what we call what happens in the corral, because let's be honest, gathering is at least a little bit fun.

Roping and draggin' and all that excitingness is still done when it is needed, but give me a hydraulic shoot, a hot shot, and an electric branding iron---when at all possible. The work that would take a dozen cowboys and the food that all of them will eat, can be done with me and my little sister with modern equipment. It's more hypnotic than the chaos-rodeo it was once. Things go so easy it's not even funny---or nearly as much fun.

But I've got all my fingers and I haven't been kicked in the head by a hoof. So I think I'll get my exciting after the cattle are worked.

I hear that skydiving is fun.

Audra Brown says that riding a horse is like riding a bike, but a horse still works in sand.

FACES IN THE HERD

I don't know about you, but a bandit-face just looks good on a cow. To be honest, I probably know more about how I like my cows to look---and can better describe---the many types of faces, coloring, and body features of bovines than humans.

Might even have more confidence telling the bulls from the heifers too.

Bandit-faced cattle look just about like you'd expect. The hair on their face is colored such that it look like they either want to be Zorro or went a couple rounds with a boxer they couldn't block.

My mama thinks they look sweet.

I'm partial to motley-faces. They come in a variety of shades and patterns, but are defined as a generally solid colored bovine with a white face that is mottled by spots, blobs, and/or specks of contrasting color. More longhorn in the breeding tends to lend finer, smaller spots, where other breeding might give a blobbier look. What's really fun is when they have a spot on their face that just can't help but look like something in particular.

Finding shapes in a motley-face's face is kinda like looking for 'em in the clouds---but you're looking at cows.

And when you find such an identifying shape, the lucky cow often ends up with a name instead of just a number on her ear-tag. I've got one right now, name of F-Hole, like the s-curvy openings in a fiddle. Beats calling her Fiddle-Face. She's the daughter of Spec, who looks like you think, and the granddaughter of Mot, who was my first motley-faced cow and gave me a long line of good stock with interesting faces.

There's solids and ring-eyes, rednecks and eyebrows, pig-noses, blazes, dish-faced, black-nosed, and monkey-faced cows. A mustache calf is never a heifer but always a star. And then you the ones that got them bug-eyes that are just hard to stare down. Accidents and bad-winters sometimes cause more unnatural features, and that's where you get bob-ears, bob-tails, and that one-eared cow.

No matter the markings or how that makes you feel, the real wisdom to remember is that the important part about a cow's face is how far it is off the ground.

High-headed is rarely a sign of a mild-tempered cow.

Oh, and don't forget Greta, bless her heart. She's a good cow, if a little yellow, and has just a bit of a crooked mouth.

Audra Brown is pretty level-headed most of the time.

THE ART OF PUNCHING COWS
"WAY OF THE WAVING HAND"

Let us look today at some of the flashier disciples of the hidden art of cow-punching known as Kow-rate. These are the men and women who confront the livestock directly, they who truly punch the cows.

Herd-animals must be maneuvered and convinced to go certain places at the behest of the cattleman that they belong to. But, you cannot ask a cow to go through a gate or walk down an alley. Rather, using the Kow-rate method of one's presence, they must be physically persuaded to move. This is accomplished in a manner both direct and yet infinitely subtle.

The greater masters of this practice reach a level of competency where the details and nuances of their manipulation of the bovine mind is instinctive and, even to themselves, unexplainable. Those great masters can quiet a bawling calf with a look and sort a pen of cattle alone. To see it, is to believe that they can move beef with the power of their focused thoughts…

While a true master is skilled in all the ways, in the beginning, a student must choose an initial branch of the path to concentrate on, and a practitioner will always hold one as their most favored discipline. There are many paths that a student can choose to take, and this is one.

The Way of the Waving Hand requires a soul both daring and sensitive. They will enter a pen of cattle with no weapon or wand but the appendages naturally given.

They have the advantage of increased subtlety and calm, but the disadvantage of distance. This is the hardest of the three paths to learn first and is recommended as a first discipline only to those who demonstrate the innate power of animal-empathy and instinctual-influence.

There is something to be learned about a man or woman by which Way they seem drawn to and most talented.

The ones who first walk the path of the Waving-Hand are the most stable and trustworthy of individuals, whose inner peace gives them great, but invisible power over the herd.

In a pinch, when they must grasp at a greater raw power and reach in order to get the day's job done, they may extend themselves with a flag or a flexible stick, an extended rattle, or even the weapon of another path, but even then, they will only wave and gesture in the gentle way that they have mastered and learned, even as they may hold in their hand the weapon of another path and more aggressive purpose.

Audra Brown will tell you more about her preferred path next, on the other side of this sheet of paper.

THE ART OF PUNCHING COWS
"WAY OF THE LONG POKER"

We're back again to peel back the mysteries of the art of punching cows.

The Way of the Long Poker is the most commonly practiced and is the most easily learned. At its core, is the simple concept of extending one's normal appendage-influence by three to five feet with a stiff cane, stick, piece of pipe, or other similar length of something suitable. The weapon of this Way, that is formally known and referred to as, a 'poker.'

There is little mystery to the use of the poker, it is direct and obviously straight. It is often applied as a broad wave of physical presence to one side or the other, or as a stab of confrontational presence directly into the body of a particular, nearby, member of the herd. It is effective, not too calm, and not too chaotic, and it is easy to learn the basics. However, while it gives the holder a great amount of direct physical influence, it is still very little defense against a cow that is intent on running you over.

Don't neglect your fence-climbing exercises.

In general, sharp, precise strikes to certain pressure points and special targets can momentarily distract or divert an agitated or incorrectly-directioned bovine.

It can also be used as an extension of gestural manipulation and as a walking assistance device after those incidents where your dodging and climbing skills proved inadequate to prevent a disabling collision.

It is the best introductory path for most, and many will remain focused on this discipline above the other techniques to eventually become the elite masters of the method. Those students are bold, direct, and disinclined to ever give any ground or take a step backwards.

They will become effective, efficient cow-punchers, confident in their own ability, yet aware of their limitations. They are unlikely to attempt to great a feat of power and will always get the job done. It is neither a tool of retreat or aggression. It is not glamorous or mysterious. It is the most balanced and reliable of all the paths. The road to mastering this Way is long and straight. A skilled practitioner is always welcome.

On a very practical level, it is, next to the Waving Hand, (which needs no external tool), the least requiring of a specialized weapon. A stick, a stout kosha weed stem, a re-bar post, an entrenching tool, or any other similar linear extension that you can hold in your hand will work as a poker and effectively move the herd in the hand of one who knows the Way of the Poker.

Audra Brown has poked a mad cow with a tree branch and won.

THE ART OF PUNCHING COWS "WAY OF THE QUICK WRIST"

Not all cow-punchers favor a direct and safer method of maneuvering bovines. For those who are ready to trade a little predictability and comfort, for the wiles of a flexible weapon, then exploration of The Way of the Quick Wrist is probably a good choice.

Known for its pains as often as its victories, it is not a path to glory, but it is a way that rarely gets boring and never runs out of room to improve.

It is the most versatile and the most difficult to tame, as it involves the mastery of flexible weapons such as the whip or a chain. It is neither direct nor forgiving. The tool is as inclined to betray as it is to be directed, and unscathed is not a likely description of the practiced practitioner, and is assured for an amateur.

The flexible weapon, may be used to extend the practitioner's influence by waving it through the air, creating a visual impression of imposing space several feet our from their body in many directions. It can also command and direct by creating audible cracks and pops to influence non-visually, and beyond the range of a physical connection.

Above all the others, this Way stands out for the breadth of its effectiveness. But power is thin, to spread so far.

If you are willing to reap the consequences of provoking a bovine---or, also likely, a fellow cow-puncher---you can snap and strike with the weapon. As it is not so much injurious as it is just irritating, the sting of the quick wrist is fleeting and fickle. Sharp, sudden irritation can be used as a successful tactic to tweak the motions of the herd. But the potential for dramatic table-turning situations is pretty obvious.

It has the advantage of reach and direction, able to influence a group of cattle that is too far to touch with anything but the whip's sharp words. It also stands apart for using sound as manipulation. It is fast, forward, and fundamentally aggressive---and thus, tends to create a fast-moving, dangerous mood among the cattle. Yet the weakness of this Way, is that it is ill-suited to physically defend against direct, agitated, bovine aggression that it inevitably engenders..

The flexible weapon is a fickle tool, likely to strike the one who wields it at the slightest chance. This is truly a path that can give one the greatest powers over the herd, but at the cost of chaos and risk of harm. It is suited as a first path for those with little fear and great ambition. But beware them as well, for it feeds the chaos in their soul and can turn them callous to the pain of others and grandiose in their quest for ever greater manipulative power over the herd—and over the world…

Audra Brown is perhaps unduly inclined towards this particular Way.

THE ART OF PUNCHING COWS "WAY OF THE FORKED WAND"

To conclude our study of the many paths on the road to becoming a master puncher of cows, let us look at the most modern of the practices.

With the invention and spread of controlled electricity, many things changed and many technologies emerged to make the world a more shocking place. Among the eventual creations is the tool that a cowboy of daring and impatient character is bound to study and use.

The Way of the Forked Wand is not a subtle art, but it is a complex and precise pursuit.

Unsuitable for much waving or gentle motions, the forked wand must be targeted and applied with specific and accurate intent. Timing and location are very important and a student of the Forked Wand must also be a student of not only the bovine psyche, but the bovine anatomy as well. And, of course, how a stimulation of a particular anatomical location is likely to influence said psyche of said cow that the holder of the forked wand wants to move.

The actual use of the bifurcated tool requires a steady and controlled hand and a well-timed thumb to run the activation button.

A good practitioner will know when to use the juice, and when to refrain; where to apply directly, and when to deliver a glancing buzz.

But there is also a cost.

The wielder of the Forked Wand will know the strength and the pain of his own weapon well. For it is impossible to perfectly contain and direct the power that it possesses. One *will* be bit by the weapon one uses, and one *will* apply the bite to the wrong human (on accident, of course). To wield the wand is always dangerous. The hand that holds the hot-shot will inevitably get stung.

The men and women who work this way are brave and also foolish. They will be found in the tightest spaces, punching the most disagreeable cows. They are there to get the job done, without making a show of it, or taking it slow or easy. Don't get in between them and the herd.

It is truly the weapon of those with the will and the nerve to get the job done no matter the obstacles, danger, or personal cost. You will find them behind the working shoots, in the squeezing alleys of corrals, and putting cattle on and off trucks while in range of many dastardly hoof cannons and horns.

Audra Brown is not a master of this art by any means, but can hold her own in a tight corral.

YOU KNOW, A BAD DAY

There was that time we had a couple of near record hot days in a row...

Murphy decided to stop by, of course, and so on the two hottest days of the year, the water tank runs dry.

The cattle ain't got nothing to drink.

To make it worse, 'stead of gettin' out and breaking down the fence to get to the neighbors water, they have a bizarre attack of restraint and decide to stay in and be thirsty.

When you realize what's going on, they've been dry for a couple of scorchers. So guess what? They're too far from the normal water, which we fixed, so the only way to get'em healthy enough to herd is to haul water.

Don't sound too bad right? But water-starved bovines ain't rational by a far piece. They get dehydrated and most of 'em just get stupid. They wander around kinda confused and weak. The others go nuts. The kinda nuts where when they spot you across an open field and you haven't even bothered 'em yet, they decide to kill you for no apparent reason.

You know, one of those days, when you have to fight off three homicidal cows.

Keep in mind that cows can get mean, but they're predictable, even rational about their behavior. Or at least, they usually are...

We stopped our two-wheeled horses a hundred yards back, just surveying the situation. Then, from the far side of the herd, this black cow came charging. She had them crazy eyes and I could tell she intended to run somebody over.

My brother popped the clutch and got out of there, but I wasn't so lucky. Because there was nothing else to do, I just planted my foot and got as stable as I could. Luckily, instead of hittin' my cycle head-on, at the last instant she ducked around the side and tried to take out the back end...flank me, per se.

Well, I had my right leg up and when she got close enough, I let'er have a nice shin turnin' kick to her ugly nose. I musta hit 'er hard, 'cause she backed off long enough for me to get myself and my cycle outta there."

And that was just cow number one.

Audra Brown mostly just kicks cows and boards.

THE BAD-NEWS BULGE

It was a green and growing wheat field. The cattle were fat and happy. But all was not as it should be.

Wheat cattle never let up. If you finally get them through the dying of pneumonia stage and into January, they can move into the next phase of dropping dead.

Bloat.

Now, bloat is hard to diagnose in a population of good, increasingly fat wheat calves. The best eaters look round just because they are that fat---and that's a good thing. But there's the good, solid, yet gelatinous roundness of belly that you want to see, and then there's the tight, asymmetrical, bad-news bulge that indicates a severe ailment related to an excess of gastric pressure.

You think you see it and dismiss it as your eyes playing tricks and your nose taking a stand, but as many times as you almost see it, you can't miss it when it's really there.

No excuses when it's higher than their back.

Fat, you see, is heavy and obeys gravity, generally sagging if it can. Gas, in contrast, does not go down so nicely.

Now, onto the nitty, gritty, smelly solutions. There isn't a good one, trust me, but there are a few equally shower-desire-creating options.

Non-emergent cases are those where you have time, help, and facilities to get the puffed-up patient into the corrals or other location where a chute or gate can be used to make the patient stay put. In those cases, upon patient capture, the garden-hose method is employed.

If you don't keep a six-foot cut of garden hose around, you've never had to deflate a bloated calf. I call it a Garden-scope, even though you can't see anything. Best practice is to coat it in cooking oil and start threading it in behind the teeth and over the tongue. Go gentle, make sure you take the proper tube when options are felt, and I suggest keeping the outside end as far from your face as you can.

You know it when you get it to the right spot and that noise and smell is as rank and unforgettable as you might imagine—and probably a little worse.

The calf goes down, the garden hose goes back in the tool box, and you try not to find anymore that day.

Audra Brown is a certified bovine-deflation expert, a skill to have, but hope to never need.

A TIME FOR LOADING

Spring is here. Don't argue temperatures with me. There have been enough warm days and enough windy days to officially call it. And if that isn't enough evidence for you, the kicker is that it is shipping time here on the high plains.

Shipping time is the best of times in the winter wheat cattle business. It means that the calves are not all dead, the wheat isn't quite all gone, and you've hopefully found a buyer for all that beef on the hoof. It's not an easy day, this day of bovine departure, but it is a day that you are more than happy get done.

It starts the moment you know when the trucks are going to be coming. Call your friends and family with pickups and proper experience and let them know at what early morning time and place the gathering will commence.

Next step, you consolidate the locations of the calves.

A little low-impact re-arranging to get as many calves, in as few pastures, as near corrals, as accessible by semi-trucks, as possible. You make sure the corrals are ready and equipped to hold the calves and to load them on trucks. If relevant, you balance the scales. If needed, you grade the roads and fix or create a place where the trucks will be able to turn around.

Then, bright and early, you head to the place where it'll all begin. The sun probably isn't up, but it's finally shipping day.

A convoy of pickups converge and report, ready to gather and sort and load, in return for coffee, scones, and a similar favor sometime. They scatter to the edges and back side of the pasture to get all the calves found and soon come together in a formation that looks practically planned out. But no formal training was actually involved, just years of knowing how this thing usually works.

The moving wall of pickups moves the herd of calves to the corrals, and more often than you might imagine, the calves just go right on in. Plenty of times, though, there's a break in the herd. A mad chase ensues by those most inclined, and they eventually bring the strayed calves back in, albeit a bit more exercised.

Then they're sorted and put into the groups of the numbers that the truck-drivers will load. Then they're gone and that always feels good.

Finally, the coffee and breakfast and conversation begins for all but the one who call this all together and to whom the calves recently belonged. That coffee will be to-go, for its off to the bank to pay off the note.

Audra Brown is one of the most inclined.

PILES AND PILES OF MANURE

There's this thing, that people do, that I find curious. I believe that it is most commonly referred to as composting.

Folks apparently build or buy a rig that ferments the garbage. It looks to me like not enough fertilizer for quite a bit of work, but I see how it might be fine for smaller scale operations.

Out on the flatland farm where I grew up, things like that aren't hardly worth counting if you have to measure it in any unit smaller than a truck-load.

To be fair, there are some methods of adding nutrients to the soil that are a bit more dainty, like liquid through the sprinkler and commercial pellet application.

But one of the best and certainly one of the more scented fertilizers is what we call 'barnyard' if we're being delicate. Just call it manure if we're being blunt, and other things if you're in a particularly bad mood, I guess.

The bovine digestive system is a large, multi-stage, process that, in addition to perpetuating the living state of the animal, produces a significant amount of green-tinted waste. This substance is, at the onset, green, smelly, and generally kinda slimy---you can call it gross for short---but it is also quite an excellent soil additive.

It's odd to think about how many miles manure often gets to travel.

First, it gets scooped out of the pen by a person with--- hopefully, a closed cab and A/C-equipped---front-end loader. Next, it leaves the pile via loader and is put on a very special truck. It is whisked away down the road at a speed almost always nearing dangerous. Quickly delivered to the farm that will be its last resting place.

If the fields are ready, then it goes straight to work, getting slung over the ground by the special flinging devices on the rear of the truck. If it is not the season, then it gets put into a new pile where it can await its final ride when that time of year rolls around.

Until then, alternative uses might include sliding down the taller, drier piles on a cheap snowboard.

It can supply any need for a blackish-green, bubbling, shifting, awful swamp that kinda remind me of smelly lava only not quite as hot. Maybe for a movie shot?

And if you need to trap some dumb cows in the equivalent of large-scale glue, it can to that as well. But here's hoping that you don't get that desperate for distraction.

Audra Brown often prefers less smelly ways to have fun.

GETTING COWS TO PUT

What do you need to be a rancher? Land covered in something that the livestock will eat, a way or a well that will keep them in water, fences to keep them off the neighbors, a hat to put your feathers in, and quite importantly…cows.

Those four-hooved, quad-stomached, moo-making, rascally bovines are kinda the core of a cattle operation. But since you can't have the cows before you have a place to put them, and you usually have to pay of the place to put them with them, finding cows to put on your putting place is really important.

However, getting them can be a bit of an adventure.

One option is to make your way to the nearest sale-barn that sells the sort of cattle you need---a word with your loan officer one the way is implied.

You sit there all day, watching hundreds of not-the-cows-you-are-looking-for go through the ring and trying not to scratch your nose or swat one of the many flies at an inopportune moment. You might come home with a grab bag of speckled heifers, older pairs, and two of those short-legged Oreo cows that might actually be aliens. All of them are tired, and have recently been exposed to every common contagion in the industry.

And worst of all, you probably ended up with more than a trailer-load, but not enough to justify bringing the semi-truck. And it's late and your hungry now, but you've got to go back after that second load now.

A better solution, albeit one that take a little more independent work, is to contact your contacts and start a cows-for-sale search. You have to be careful, because you don't want to end up with cows maladjusted to the climate and conditions that they'll soon be living within.

You want them from relatively nearby, but not so close that they can just jump the fence and go back to where they thought that they belonged until recently. You may have to plot some midnight transportation to get them all back and you might've had to twist a few arms to get them to sell them at all, because ranchers don't like to part with good cows. That is just a cold hard fact.

You probably won't get as many as you really need, but be glad you got the good ones that you did.

Once you've got some, you can be lazy next year if you want and just keep the heifers instead of looking elsewhere. Eventually, maybe someday, you can be the one who the cow-buyer has to talk some cows out of,

Audra Brown never has enough cows.

GIFTS FOR THE AGRICULTUR-ALLY-INVOLVED

The end is near yet again and if you are impressively organized, you've got your Christmas shopping done and have nothing to do but enjoy the upcoming couple of weeks of time and food and family.

However, if you're more like me and get distracted and delayed more than a little, there's still a lot of decisions to make and things to get and do. As I've done before, I'll do again, and so here's some thoughts on what to get the agriculturally-involved associate for Christmas.

Since they've already got a fifteen-inch adjustable wrench, a portable espresso machine, and an electric grease-gun due to our previous discussions on the subject, how about something a bit less work-related?

While a new pair of fencing pliers is always good because it is the only thing that might make one look forward, at least for a moment, to building and repairing the wire-strung menace that sometimes succeeds in separating livestock, what about the more recreationally-inclined? Many, especially the younger breed, enjoy a little distraction.

For these modern souls, who have an endeavor other than the work, a likely option is a game camera. These remote sensing devices are fun and useful.

Park one somewhere on your daily feed route, or the path you take to check all the sprinklers and wells. You never know what it will have seen while you were elsewhere. You are bound to catch images of varmints, critters, and trespassers---and since two out of three of those can be worth hunting, you've got something to look forward to with your friends when you get a little time for sport.

Another option is to get someone a little unnecessary style. How about a holster for their favorite tool? If you got them the pliers last year, get them a handy holster so that they can carry it around and show it off.

Did they overly injure their mobile communication device? Did it slip out of their pocket too many times? Maybe a more secure, more protective container is just the thing.

And if they really do have it all, then let me propose that I know a thing that is extremely new. There are some books written by a local author and ag-adventurer that you might know from her columns in this very newspaper. I guarantee that not everybody has one of them yet.

Audra Brown knows what her people are getting to read this year.

THE TINKERER'S EYE

The weather's getting a little cooler at night and just in time too! It's Fair Time and that's good. The fair is many different things to many people, but whether you're there for the lambs, the chow, the music, or any one of the too numerous to list events and shows, you'll have hard time failing to see the big, bright, collection of carnival rides that is parked just off to the side.

It's a sight to see, of that there is no doubt. Flashing lights, spinning contraptions, raucous sounds, and that carnival music coming from all directions in a rather non-harmonious, but unmistakable music. Do you like to ride the rides? I'm all for folks enjoying themselves, but I'm also a proponent of knowing what you're getting into. That's my creed. Always know the risks. Don't let the danger surprise you, be expecting it. And if I die, I hope I knew it was a possibility.

With the idea of information in mind, this all reminds me of my personal experiences getting to go ride the rides at the fair.

As a young'un, those rides look like Las Vegas after sunset, Six Flags, and Disneyland all put together into an exciting, mysterious adventure that you just can't wait to explore. That's not a bad perspective, but what I learned to be true, was that it tends to be put together with duct tape and Elmer's glue.

My first rides were simple. The carousel and the amazing Ferris wheel. It was fun, and always over too soon, but I looked forward to the next year. Then I was an experienced double-digit and convinced my folks to let me go with some acquaintances who were the lucky sort who got the armbands that allowed you to ride as many and as much as you wanted. It was a young'un's dream come true, and it was pretty fun, but it was also the last time, as I found out that my years of farm fabrication and repair, had jaded my vision and I saw way too much of what wasn't there.

The stress on that structure, the suspiciously unfilled bolt holes, the duct tape brackets, and the broken light bulbs. I saw it all that time, too clear to ignore, and I was happy to let my friends ride some rides while I stood and watched from the ground.

Turned out I wasn't paranoid. The ride I went on twice, went a lot different the second time, and I realized it had been a broken run that I'd earlier survived.

There's lot of things that my reluctance might be mistaken for, but risk is always part of the equation with fun on the other side, and the math just doesn't quite add up when you're looking too accurately at those carnival rides.

Audra Brown enjoyed it while she enjoyed it, but an engineering-inclined mind is eventually disillusioned.

MY LITTLE TRACTORS

The John Deere house is a special place when you grow up on a farm that runs green tractors.

You go there to get parts, visit, occasionally consider buying a new piece of equipment, and if you're a good kid---to get toys and caps.

Ertl toy equipment is the Holy Grail. Detailed metal replicas of tractors both new and old, combines with both headers you understand and headers you don't, and working implements to pull through the dirt on your very own imaginary field in the yard (or on the turn-row).

These miniature marvels were amazing to a kid and you just couldn't get enough of them. The detail and realism was so well done, that you could practically see the exhaust coming out. Unfortunately for all involved, the price was also realistically scaled. Expensive full-sized equipment was high, and scaled replicas were similarly not cheap.

This was no deterrent for you as you spent every extra moment during a visit to the John Deer house (or other equipment dealer) lovingly examining the toy equipment selection and not so subtly pointing out just which pieces you really needed to your parents.

Your collection probably got more time and caused more arguments among siblings than any other toy that didn't have a real engine in it. You had to share (reluctantly) your implements and let little sister use one of your tractors, and try to figure out how to haul your dirt to the elevator in the front yard without a toy truck.

Come to think of it, playing 1:32 scale farmer is a lot like the real thing...

They were also useful as alternatives to other common toys. Who needs dolls when you have tractors? The big ones can be the parents, the little ones can be the kids, and the Caterpillar bulldozer can be your favorite.

Line them up and introduce them to you friends and family. Lift their headers over the side of your plate so that they can have some breakfast too. Tuck them in next to you in bed so that the can be there and ready to play in the morning.

Toy tractors are important people. You can never have too many or play with them too much. You can only grow up and wish you still had the time to pretend to be a farmer.

Audra Brown likes to pretend to farm.

THE JEALOUSY OF OLD TOOLS

There's a tool for every job and a there's also a lot of tools that will do the trick with a little bit of determination and ingenuity---cleverly cornered into something far from their original intended purpose.

The list is long of things that will do instead of a hammer.

The list is short of things that can't be done with tie-wire, duct tape, nylon ties, and a large adjustable wrench.

(And just for your information, the latter set of supplies there can make many hammer-like solutions.)

A person's tools are more important and personal than they might at first appear, particularly (though not exclusively) if that person is in the agricultural realm.

People will ask why you need so many tools and of course, why do you need another when you already have more than more than one?

Let me attempt to elaborate on a couple of the reasons that tools are important. Old tools are special. New tools are usually a mixed bag of good and bad, but we love them anyway. Any tool is good for something and it's nice to have at least one of each---just in case.

Imagine you are miles from anyone or any supplies other than those tools that live in the vehicle you are trying to fix. If you are lucky, you drove your own pickup to the field and are so equipped to the degree that you have chosen to be. Alternatively, you might have just moved from a distant field or gotten a ride to the tractor and so are limited to what you can find under the floor mats and in any small tool-storage compartments that might or might not exist on the equipment you are running.

If you do have exactly all the tools you need to fix the problem, it's probably a bad sign that the same problem occurred on the same rig in the recent past. We all know that tools are by nature, vagabonds. The personal set stays put the most, but only due to consistent supervision and dire threats of consequence should a tool become AWOL. But those that do not serve a specific mechanic are a fickle and wandering lot.

There is a very particular excitement that comes with getting a shiny new do-something-a-ma-jigger. But beware, both the old and faithful makeshift tools and the other tool-lovers will be jealous of the newcomer.

New tools are an addition that's always welcome. But they must be watched. Keep it close. Keep it safe. For some strange reason, new tools always seem most likely to R-U-N-O-F-T.

Audra Brown wouldn't say no to a new tool.

KINDA SORTA CAR-HOPPING

We live in a wide land where automobile transportation is not a benefit but a necessity.

Rules for dealing with cars are almost as important as ways of dealing with people. And dealing with people's cars and people in cars, is pretty much the epitome of need-to-know.

Getting into someone's vehicle without permission or invitation is a pretty serious line, in general. And I certainly don't recommend ignoring that important standard of conduct in your everyday life.

But, as with many customs, things can be a little different on the farm.

I say you never need an invitation to jump on somebody's tractor to bring them lunch, a snack, or something cold to quench their thirst. But then again, you might say that stopping to not run over you is an invitation that is clear enough.

The only time this would be real iffy is when the thing that they are doing to the crop is one that must be done a speeds slow enough to safely hop on and off. Say, they are running a peanut-thrasher, slowly, down rows that will hold the tractor on course...

I'd recommend waiting until they hop off to check
around the back and be sitting there when they return.
Though, be sure to have the snack clearly visible
to prevent sudden retaliation in response to such a
surprise. And be ready to give them a hand so that they
don't jump the wrong way and fall off.

When you're not the one delivering sustenance, then the
thing to keep in mind when looking for an automobile
to invade, is knowing who is likely to have the best
supplies, a working temperature-changer and a back seat
that is empty enough for you to find yourself a space.

When it's a cold day at the farm auction, the protocol is
at it's most relaxed, but also in demand. Nobody cares
who's in the driver's seat of the pickup that is nearest,
you only need to look for a head in the seat that you are
planning to soon occupy, and jump in the door where
you can't see anybody's noggin.

Open the door, close it quick to be polite, and if you
picked the right rig, you'll find friends, maybe some
egg nog that includes the rum, and at least one sharable
liquid-holding device.

*Audra Brown has had good luck hopping in the best
stocked pickup.*

BUZZ OF THE GENUINE ARTICLE

There's a change in the air. As the days get more regularly warm and the random snow-days abate, people find themselves out and about, in the out-of-doors. It's time for listening. It is the time to keep your ears open and your boot-tops high.

There be snakes in the sand.

That tell-tale rattle is something that you can never hear without at least a few hairs going up on the back of your neck. And though we all get a little antsy and worried about those sounds that sound similar, it is much easier to mistake a snake-like rattle for the real thing than it is to mistake an actual buzzing snake for some impostor. There's just something about that noise. You know it when you hear the real thing.

No questioning the genuine article.

You are already wondering where you put the nearest shovel and if you are packing iron today. If neither of those obvious solutions seem close enough, then you start looking for anything and everything that might serve as a remedy for a rattling snake.

Things that you can throw are good if you have a decent tossing aim. But better not, if you can't hit, because just getting it mad is not so good of an idea.

Pin it with something long enough to make you feel safe,
you can finalize the cure with your pocket-knife.

The knobby end of a rope or booster cables can make for
a powerful and effective swing, but again, better be sure
of you aim and remember not to jerk those hook-like
ends back towards your face.

If all these seem too easy and old hat, then make it a
numbers challenge. Dare a friend and fellow snake-
doctor to a summer-long contest and don't ever let them
forget when you are ahead.

If you're looking for tricks to beat, there are stories that
are hard to believe, but true nonetheless. Knock is out
with the force of a bullet passing by, then finish it off
with a rock or a knife. Remove the head with a fast-
moving lead treatment but at an angle such that it flies
into the air. Then catch it as it falls if you are sure enough
in your precision to assume that it won't have a head.

Or just take a shovel to it in the classic way.

Audra Brown doesn't have any points yet this year.

SHIPPING DAY

What comes around once a year and is always an occasion for food, family, friends, and more than a little celebration?

Shipping day.

You look forward to it for months, just waiting for the day when you can trade a few truckloads of troublesome adolescent bovines for a check that you can spend buying more of the same for next year. In preparation, call your friends, family, and neighbors who have the knowhow and the vehicles to help round up a few fields of frisky cattle. Call the brand inspector, make sure the buyer and his trucks know where and when to show up. Find out if Mom or Grandma is bringing the biscuits and coffee.

When the morning comes, everybody shows up at the corrals bright and early, converging from many directions through different gates. Vehicles range from farm auction fodder to the pickup that usually is reserved for pulling campers or picking up groceries in town. Four-wheelers, dirtbikes, and even a Deuce-and-a-Half have shown up too. Every man, woman, and child that can drive brings a rig.

Carpooling is neither an option, nor polite.

If you have to borrow a pickup to have something to drive, it can be arranged. Drivers are always more scarce.

Once all vehicles are manned and the gates are open, we swarm out onto the fields in a ground-covering formation. Cattle are pushed from the farthest weedy corners (though not by the town rigs, who stay on the open field) and the bunches are brought together and headed towards the destination corrals. It's quite the sight to see all those pickups spread out in perfect coordination. You'd almost swear we'd practiced before.

Don't though, because we didn't.

If all goes well, then all the cattle go in the pen the first time. If all don't, then the fun times are on. When coordination and numbers fail, speed, experience, and a confrontational relationship with apprehension prevail. While a few wiser souls start sorting and loading the first truck, the rest of us go back out and show the crazy cattle who's crazier.

Eventually, all but the cuts are on the trucks, the happy former owner is on the way to the bank, and the rest of us are sitting around eating biscuits, scones, cinnamon rolls, and hopefully, apple burritos---washed down with coffee, tea, and coffee. It's a party for a little while before everyone has to get back to their own business. No one doesn't look forward to shipping day.

Audra Brown has never been one of the wise ones.

DUST OR MUD

What is the worst part of standing in an alley with thousand-pound cattle running head-down in your direction?

It's not the hoof-cannons they fire in your direction as they run by.

It's not the nerve it takes to stand your ground and hope that attitude beats size when you need them to turn around or veer another way.

It's the dust.

The dust is the problem that you can't avoid.

It gets in your eyes and sticks to the front of your teeth. It works its way into your gloves and under your clothes. It obscures your view and makes it even more difficult to tell the difference between a bull and a heifer and your dad.

There's not much you can do about it, even if you try.

You can get a water hose and spray and make the top layer seem a bit better, but a few hooves later it'll be stirred up so much that you can't tell the difference. You can flood the pens and deal with mud instead of dust…

But slipping and sliding yourself is bad enough without the added excitement of thousands of pounds of beef losing control as well.

There's really not good way to fix it on your own, but every now and then, the weather decides to be on your side and you get to sort cattle in dirt that is perfectly wet.

Perfect corral condition.

It doesn't get in your face or in your nose. You can see the whole alley and don't have to guess what is headed your way.

It's almost unreal how much more pleasant it is when the moisture to dirt content is the optimal ratio. I've got to write it down or I'm certain that I won't believe that I had the experience.

Who knows when it will happen again. Perhaps it will never be so nice on the exact day that the sorting needs done, but I will always remember what perfectly moisturized corrals is like and hope to feel that way at least once more.

Audra Brown is not as dusty as she normally would have been.

AG-SHOWS OF ANY SIZE

As I'm sure we've covered before, a good reason to get off the tractor and go somewhere is always appreciated...

...and the most rare is that sort of opportunity when it isn't due to something breaking down and requiring a trip to town for parts.

The county fair, the annual board meeting of the bank that lets you borrow the money to keep farming, the annual meeting of the power coop that gets quite a bit of that money, the cousin's graduation that only happens once... There's a sort list of not-often events, but one of the big ones has got to be the farm and ranch shows. There's the one far away that doesn't always get attended, and then there's the one close to home.

Not all ag shows are big, but that's not a problem. Sometimes it's the quantity of cool tools and equipment. Sometimes it's the concentration of people worthy of a long and rare conversation.

Climb on the new tractors and dream about working in such a clean cab with all the space-ship like controls and A/C that actually works.

I'm captain of the field in my mind in those things.

You can admire grain carts that haven't yet smelled

like rotten wheat, flatbed trailers that don't have even a scratch on the paint, and cattle trailers that are still the color they were painted, not the shades of green that the cattle squirt all over the floor, the roof, and the walls.

There are booths and peanuts and candy and hopefully yardsticks to carry around.

But the best part is the people that you'll run into or that track you down.

When I was closer to a yardstick in height, I'd wait for what seemed like hours for my Dad to talk to someone about something.

Back then, I didn't quite understand the allure of a good conversation, but now it makes all the sense in the world and you'll find me doing the same thing for probably the same reasons.

People you know, people you ought to know, people you'll get to know anyway, and people that you probably do know, but for some reason, you just can't quite remember.

So, come out and see the shiny equipment, talk to the interesting people, and take a day off to have some fun.

Audra Brown hopes to spend too much time talking to you unless there's a combine to pretend to fly.

MIDNIGHT MEALS

Do you know how hard it is to find a decent meal at two o'clock in the morning? In a small town?

Big cities seem to have both a reason and a willingness to do business after plenty of folks have presumably retired for the evening. Most times of the year, us country folk aren't anywhere near town at that time of day. Or any time of day, really. We're not there much, but when we are, midnight is nearly as likely as noon.

An event that guarantees this moonlit behavior is that autumn marathon known as Peanut Harvest.

It happens in the fall and once those vines are dug and dry enough to thresh, the harvest is begun. You run the threshers just as many hours a day as you can. Sitting on the tractor, trying not to forget to lift the header on the turns, and otherwise occupied with keeping the thing running---that's stressful, requires experience, and is probably the primo job.

Then there's the kid who's on the "entry-level" position. The catch wagon.

You have never seen so much work with so few moving parts. A tractor, an army of empty trailers, and if you're lucky, a chain or a friend to help lift the tongues.

Hook up an empty trailer, catch the threshers when they
get a bin-full, drop the full ones in a line by the road,
repeat until harvest is over or you lose against an evil,
heavy, reverse-spring-loaded trailer tongue.

Both these jobs are localized and don't end up in town.
But that's where the peanuts need to be, and guess who
gets 'em there? Yep, that's the job that ends you up in
town at 2am with hypoglycemia and a smashed thumb.

Dragging peanut trailers is the job you graduate to
that makes running the catch-wagon seem like a nice,
simple, yet exciting job. By now, you're old enough to
drive to town, big enough to lift tongues without 3-point
assistance, and if you aren't already, you'll be a master
of backing a rig with two points of articulation up to a
trailer you can't see and hooking 'em together so that
you now have a rig with four.

Drag 'em to town at a speed reminiscent of a tractor,
hope you don't break an axle this trip (cause you will
eventually), and still have a good couple of hours of
work even after all the tractor-drivers have gone to bed.

Supper at seven is a long time ago at two in the morning.

Audra Brown did find some hamburgers
and ate all of them.

NOT OFTEN SEEN ON LAND

Spring on the high plains. Time for sunny days, shipping cattle off of wheat pasture, and straight-line winds not normally seen on land.

Conventional expectations are fine, but there are more than a few things that defy the definition you'll usually find in the encyclopedia.

Amateur wind speed identification charts that I've seen usually stop at 50 miles per hour with a note that you just don't see that on land except in a circular storm of the hurricane or cyclone variety.

This leads me to conclude that we do indeed have the market on horizontal gustiness.

Another somewhat impractical guide that I recall is the official instructions on what to do, as a motor-vehicle driver, when the air take on that sandy composition.

I've only encountered a couple of yankee-chasers where visibility was bad enough to consider pulling over and letting it blow itself out.

There a lot of time that would have been spent not getting things done if we pulled over every time a dust storm came along that blocked some of the sunlight and cut the visibility down.

An amusing misconception that is sometimes held
by those who haven't been to this part of the windy
world, is that tumbleweeds are an invented accessory
for old westerns and not an impossible to exterminate
menace that can cause more than minor inconvenience
and pain. It is more than surprising the first time you
see tumbleweeds treated as an amazing mythological
experience rather than a bane that ranks up there with
grass-burrs and bind-weeds.

It makes you realize though, that we do live in a place
of myth and legend. We gripe and complain and pray
for more rains, and we do need the moisture, but there's
something about this place that just can't help but be
impressively impressive.

So, haul off the trees that fell, rebuild the hot-wire fences
that departed, and go ahead and seal up that window
that you didn't realize was so loose. Then, look out and
realize how far you can see.

Look at the mountains on the far, far horizon. Go out at
night and look at the milky-way up in the sky. Realize
how quiet and still and beautiful it is on all those days
when the wind isn't blowing.

And the rest of the time---it's legendary.

Audra Brown has seen those speeds on land.

NO PLACE THAT STAYS PUT

With every job, there is an location where you are bound to spend some notable portion of your time. Some are quite thoroughly situated in a spot that we call an office. So core is the location that we even go so far as to call them "office" jobs, defining the vocation by the location.

Well, there are very few jobs on the front lines of agriculture that you can even think about calling by the office. I mean, yeah, you might spend an hour or two in the office of someone else while they load your trailer full of chemical, or they fix a set of flat tires while you wait. You spend a little time in your loan officer's office or sitting at your accountant's desk. But, even then, during tax-season, there is no place that doesn't identify as mobile, that you can say is your place of work. To define any sort of constant location is to end up with a area that would give an more easterly state and run for its money, if you needed to play that game.

But to say that there are no spaces that you spend most of you time would be an untruth. Though, unlike an office, where you go to a place to stay there for a time. These ag-offices are the place you are while you are going to and from the place you need to do something.

The pickup, the tractor cab, or, when you are being more of a rancher, maybe the back of a horse.

The saddle is perhaps the most austere of the mobile offices that you will find. It serves as its own decoration and cannot hold or collect a great deal of tools and trash. At least you chair can't be stolen away.

The tractor is odd in that it is the cubicle of the lot, with minimal space and more generically applied. You have a cab that you can fill with tools, dirt, trash, books, tapes, odd weeds that you thought were interesting, and of course, odd rocks. You might have a ukulele or other small instrument to play with, and the snack stash, ice chest, glove collection, and hat rack.

Sometimes you get assigned a cab for months to yourself, and sometimes you have to share with the day shift and your brother and probably someone else.

But if anything is yours, full of your stuff, mostly stays the same, and is where you spend the majority of your time.

It is your pickup.

A mobile HQ full of tools, parts, and cold beverages of your choice, it's the home away from home that follows you from one end of your space to the other, in-between, to town for parts, and back again.

Audra's office is a bit rough, having been run into by some other offices a few times.

FASHION FIT FOR THE WIND

When the wind blows for a day---as long as that day isn't shipping day or one of the other firmly scheduled events on the farm or the ranch---you can probably find enough inside jobs to keep you minimally sandy. But when the wind takes up residence for a longer span of time, avoiding it is just out of the question.

Advice on the wind isn't new. We've heard not to spit into it and to throw caution to it when we can. I'd also certainly posture that you ought not run around in it with too wide of a grin.

Let's see what else you might want to know about taking care of business out when the weather blows too much.

Preparation is first. Before you head out into it, you need to have your attire sorted and tied down. I prefer britches-legs down over the top of your boots and even if they are too long or you are too short, rolling up the cuffs is not the best idea if you don't want to get weighted down.

There's not much to manage with your shirt and coat. Fewer pockets is fewer pockets that will fill up with sand, but I can't really say that it is worth the loss of places to put stuff.

You'll have to weigh you own options on that one.

A mask or bandana to redirect the larger particulates from going up your nose is a pretty good preparation. It'll help reduce the mud in your mouth as well. Just remember to pull it down before you go into any banks just to keep any wild western ideas from causing people to mistakenly call the law on you---unless, of course, that was your intent.

The thing you put on top of your head is important every day, as we all know, but the choice you make on a windy day has the most consequences by far.

If you wear your hat, the brim is gonna catch some wind. If it fits down and tight, and you trust it to stay put, you'll probably be okay. But if you doubt it all that it's gonna stay stuck to your head---and like it enough that you'd really hate to lose it again---you might consider a ball cap as an alternative.

Because a hat that blows off is hard to catch and rarely seen again.

I hope this is helpful for the next time you need to narrow down your fashion choices before you head out into the flatland winds. And if you're out of inside jobs, hats off to you, and may you not need to open too many tumbleweed-buried gates.

Audra Brown is writing this inside.

CORRECT ANSWERS TO COLLECT

Problems to solve are the commonest of events on an
agricultural operation.

Many require only mental effort to figure out and plan.
The carrying out is only an afterthought once decided.
But the greatest problems are the ones that you don't see
coming and that will probably kill you if you don't come
up with the correct answer quick.

When the mad-eyed cow decided to flip your
motorcycle over, the correct answer was, in fact, a solid
turning kick to her face.

When the calf got his head stuck in the swinging gate
and you were down in the mud, getting run over and
over and over, the correct answer was to sink down
lower and wait.

When the fire is burning up the grass right up to the
house, the correct answer was to scoop it up in the skid-
loader bucket and dump it, along with the fire-retardant
sand that came with, over a not-supposed to-be-on-fire
tree---over and over again.

When the doohickey roller inside the top of the baler
needs reinstalled and you have to hold it in place while
hanging off the side quite a few feet off the ground, one-
handed, and blind...

...the correct answer is to hold on for dear life and get a massage for the over-worked arm after you pry it stiffly from the place you've had to keep it frozen for far too long to accurately remember.

When the vehicle is rolling off down the hill on its own, sometimes the answer is just to let it go and sometimes the answer is a bit more on the nose---or on the grill-guard. If, that is, you are conveniently positioned downhill from the runaway and you know---or at least suspect---that the grill-guard is firmly mounted.

When the smoke from under the hood of the pickup you are driving is more than just a coolant leak, the correct answer is to grab your phone, your gun, and the portable espresso machine...and exit door left. Then, don't look back until you've put some significant distance between you and the pile of fire that it'ss about to be.

One thing about the unexpected problems that you get the chance to answer correctly and beat---you end up with a collection of prepared answers that you really hope you'll never need again.

Audra Brown has, so far, had the correct answer to the deadly problems.

TRACTOR "SELFIES"

It has come to my attention that the "selfie" is now a popular and culturally prominent behavior. At first. I found it to be a little silly. At second, my first impression still seemed pretty solid. However, there does appear to be situations and circumstances when this is a reasonable---or at least understandable---action.

When you are alone and in an unlikely to be repeated situation of some significance or importance, the "selfie" is a poor solution. If you feel that some photographic record is necessary, it may be the only option, but that's more unlikely that you're likely to admit.

Better solutions include setting the camera up with timer or finding someone to hit the button on your behalf. Unless you are in a prison exclusively populated by kleptomaniacs, a stranger that you can trust as far as you can throw them is usually to be found.

(But be sure to keep them in that range, of course.)

When the subject that needs to be recorded in the same camera shot as you is the one other person you can find, then again you have a conundrum. I concede that there exists understandable motivation to use the technique in light of there being occassions with a distinct and factual lack of other available options.

Nonetheless, there seems to be an unnecessarily frequent engagement in "selfie" activity that is neither justfied, nor, in my opinion, a benefit to the cultural paradigm.

Boredom is a motive and could be a reasonable cause. Those long, captive, lonesome situations...can lead to an absence of anything else or better to do. I don't know if they exist enough to forgive any decent portion of the mass of selfies, but they do occur, alas, in the field.

I came to the unsettling conclusion, that I, and other farm kids, may have inadvertently and unknowingly engaged in "selfie" behavior before it had a label.

When you are in a 4' square cabin for more than half the hours of the 24-hour cycle we call a day, for weeks or months of days...you've heard every song on every station you can get, you know more about current events and politics than the pundits you heard it from, and you've become an expert on every nuance of the tractor you're living in.

In that situation, when you have a phone that takes pictures, it might seem reasonable to utilize this tool to enact a thorough study of your face.

Audra Brown has conducted some facial self-studies.

WHERE THE WEEDS GROW

It is a fascinating fact, that even in the dry, not-so-lush lands where a decent crop requires more than a little irrigation...the weeds will always grow.

The less fascinating, but more problematic truth, is that they will grow faster than the useful plants. A field is like one of those old western towns that only exist in the movies. There's just not room for both the good and the bad. (Ugly is fine.) The point is that the weeds will take over a field in a hurry if something isn't done to keep the undesirable flora out of town.

So, unless you're extremely fond of tumbleweed salad (which, mind you, is an acquired taste that no one I know has been able to acquire) the farmer needs to fight the weeds and win.

Before the crop is planted, you can plow and disc and plow again until the ground is as brown as dirt. The bonus with this method is that you get a nice tilling too. It's good to fluff and mix up the soil every so often.

On the other hand, it does no good to have fluffy ground to plant in, if it blows away first because you made the mistake of turning it loose for too long, where the wind doesn't have to work very hard to blow it to Texas.

And that's often the state of things---around these parts.

No-till is what you might call an option. Spraying is a lot faster than plowing and you have to stop and grease the equipment less. But you'll be likely to need to keep more parts onboard due to the fact that all those rubber and plastic nozzles and hoses are both fragile and seem somehow attracted to fences and such. Not to mention that no-till leads to an increase in the depth of tire-tracks and a general state of the ground being rough.

When the big boy toys are just too much, too broke down, or are for some other silly reason not an option, you can take a precise, but labor intensive approach.

Maybe you do this regularly and have a nifty spot-spraying rig that hopefully still has foam on the seats and a shade for each player. We can draw straws to see who gets to be the driver.

A bit farther down the techno-ladder, there's the jury-rigged rig that looks a lot like an ATV with a spray tank bolted on somewhere. Requires one unlucky driver.

Last but not least, the old standby that I don't even like doing in my backyard garden... It looks like a hoe and that's exactly what it is. One long-handled tool per person and you can take on that invading army of weeds, one plant at a time.

Audra Brown is not looking for a hoeing job.

WHEN IT POURS, IT RAINS MORE

When you live in a desert, rain is rare, by definition.

When you live in a desert and are involved in agriculture, that rare rain is precious and always appreciated---even on the occasions when it also causes some inconvenience. Usually when that water does fall from the sky, it is so quickly absorbed by the thirsty earth, that any perks other than just moister ground are not experienced.

But every once in a blue moon, there comes a storm that really brings a big bucket and the water flows.

And that can be a good time, especially for an ag-kid.

If you're out working, then you probably were allowed to stop early when the rain started coming down hard. You got to enjoy the rare experience of getting wet without taking a shower or falling in the water tank or unclogging a sprinkler drop. Maybe mom made some hot coffee of cocoa when you all got back home. Maybe you were in town in your not-a-pickup and had to get a friend to take you up the river that used to be a dirt road to get back to your house.

Either way, the first perk of a really good road-washer is that you get to take a little break and spend some time with the ones that you like to be around.

Then, the next bit of fun has to be caught and taken advantage of quick, before the water does soak in and things dry up. But if you hurry, and you know the right sort of friends, you can catch a morning of kayaking down the barditches before they stop running water.

If you know where the water likes to run and there's a proper sort of low spot where it eventually ends up, you can find a temporary swimming pond and practice that usually unused skill while the water is still fresh and cool and suitable fun.

When it starts to dry up into the perfect viscosity of mud, the fun isn't over if you like practicing your stunt-driving skills, and you have an available 2-wheel drive pickup.

Start with the one-eighties in a place with a safe margin of around, in case your skills are rusty since it's been a blue moon since the last practice run.

Remember, a big rain is a short vacation with more work in a day or two, so have fun, be safe, and enjoy the flood.

Audra Brown needed a boat, but still doesn't have one.

NO CLOCK IN THE BARN

There is no clock in the barn.

Or, if there is, it's probably wrong because who remembers to take care of the clock in the barn?

There is a reason that there is no clock in the barn. The barn is for things that need doing, not things that you can pause at an arbitrary moment. When you are welding up a gate or fixing the planetary gears that go inside the hubs of a rather large tractor, you can't drop what you are doing---literally or figuratively.

Indeed, it is equally as difficult to gauge how long it will take at any point before the project is complete. Inevitably, your estimate will be wrong.

That really hard job that you've been putting off for a long while? It'll be much quicker than you thought.

That simple oil change that'll only take you an hour or two? You can laugh about it in a couple of days, maybe, if you're done...

There are important times in the barn.

There is time to eat. Either the first intersection of hunger, empty hands, and nothing being on fire...or, when somebody shows up with food.

There is time to sleep, which is more difficult to pin down, but somehow, approximately once a day, you find a pause-point in the project and take a few hours to recharge. Or, more often the case, you work until your realize that you can't go any further without some part, tool, or supply that you don't have, no one you know within a few miles has, and can only be gotten at some place that isn't available until tomorrow.

But all the time spent in the barn isn't work and the lack of a timepiece is equally important in those cases.

Barn party?

Time isn't important once you arrive.

A game of pool with your sister? A movie projected up on the wall? An old car that you only work on for fun?

There are lots of fun things to do in the barn and the best part of having fun in the barn is that there is no clock telling when to stop.

Work hard. Play hard.

And don't put a clock in the barn.

Audra Brown doesn't clock-in, she just clocks-off.

BE SKEPTICAL WHEN WET

Here on the high desert, water is precious and rare.

The smell of moisture is more fine than a flower in bloom and possibly even the scent of spoon-floating coffee---though I'm not inclined to put anything above caffeine, personally.

We celebrate when it rains, smile when it snows, and even find the damp lining in sleeting, windy, blizzardy storms that are honestly hard to not find unpleasant.

But where it is not often an occurrence, a thing is bound to be looked upon with an eyebrow raised and salt ready to be shaken. Despite all the hope for good moisture that we hold, the presence of water is more often than not, an indication of something that has gone wrong.

Water in the road, running down the ruts and under your tires. Slipping and sliding and doing one-eighties like the residents of Hazzard County is fun, but the sad truth is that you've probably got a pipeline busted and better slide over to the backhoe and get digging.

I hope you have a backhoe.

Otherwise, I guess you better break out the shovel.

I'll help run the backhoe or I'll bring by a cold drink.

The sprinkler pivot sometimes looks like a water fountain, shooting artful sprays in every direction and ensuring that you are gonna walk to the control box on a pool of water that just can't help but be good at being conductive. If you don't get fried, you'll still get wet. It's one of those situations that's just gonna be unpleasant no matter what (and may cause injury or death...)

When the corrals are a green, smelly waterpark of slip 'n sliding cows, you can put on your muck boots to keep most of it off your socks, but don't be surprised when you're more green than not at the end of the day.

It's practically physics-defying where that slop will get.

But the time when you should be concerned, is on a windy, uncloudy day. You feel that drop of moisture and its friends on your face. Everything is brown---the wind is blowing, after all. But even though you can't see it, you know that yellow is the color that hit your skin.

Guess you can be glad that it's too breezy to smell.

Alas, there's nothing to do about it but complain when a cow is upwind.

Audra Brown can try to get downwind, but too often, she's just surrounded.

JUST GIVE IT AWAY

We all know that technology is obsolete before you even get to see it, most of the time. You order that new phone and there's already an upgrade before it gets to your door. This is important as a concept because there is a lot of technology on the farm. Mind you, it does tend to advance at a slightly slower rate, but it is nonetheless a mess when you realize just how useless a piece of equipment has become.

There is both pleasure and pain that comes with the obsoleting of some critical tool in the farming process. Often, and hopefully, the new version of getting the job done is somehow easier. You can't help but appreciate things that make you get stuff done faster, cheaper, more comfortably, and with less likelihood of permanent (or even temporary) maiming and/or death.

But it is painful to see a tool that you've likely invested many dollars, much time, and a great deal of energy into in order to keep it running, be prepared for its likely breakdowns, and improve it so that it does advance and keep being worth it---not be worth it anymore...

Alas, eventually, no amount of blood and sweat and modifications can prevent the inevitable state of irretrievable obsoletion and you must move on and get with the times.

If you were less busy farming, you might have had the foresight to see it coming and sell the dadgummed thing before everyone else figured out what was coming too. But no, you didn't have that extra attention to pay and so the situation arrives where it is all but worthless.

You've already realized that you don't need it anymore because you've acquired, one way or the other, its replacement. And so has everybody at the farm auction.

It's amazing how many extra bottoms you have for that old-school lister that was the cream-of-the-crop in its day. You remember how many miles you traveled to collect the full set of that weird size with the proper hardening and bolt pattern. You see the former potential and the preparations that were made for a future that it didn't have. And you see the equally clear truth that it is just no longer the way to do the job that it used to do.

Alas.

Alas.

So to the auction it goes, and out of your way. It seems like you're selling it, because that would make sense, but in all but definition, you're just giving it away.
Alas, again.

Audra Brown just likes to sit and contemplate obsolete plow bottoms.

119

TALKING OVER THE FENCE

When you live in the country, you land is more than your residence, it's your life. And as they say in the real estate business: location, location, location.

Your land not only has to have a place for you to live, but it has to support enough cattle to support you and itself.

The cattle are most of the day-to-day work, keeping 'em fed, watered, and healthy. But then there are the logistics of keeping them where they are supposed to be.

Fences are never quite good enough, and eventually, you'll get a call from the neighbor and have to go round up your strays and bring them back home.

If you're lucky, you'll notice the misplaced bovines quick, before they get too spread out and mixed up with the neighbor's herd. If not, what you told yourself would be a quick little job will turn out to be a two-day ordeal full of flat tires and angry cattle.

But the fact is, you've got to put your cows back in before they eat all the neighbor's grass.

The cattle don't realize that eating the neighbor's grass is not only inconvenient, annoying, and unnecessary, but is taking away from his life, his family, and his business.

Sometimes, it's kinda funny, but it can be a big debate as to who is in charge of keeping up the fence between two places. Neither side wants their cows to get out on the neighbor and neither side want the neighbor's cows on them.

Sometimes, a valiant altruistic sort will just volunteer and everything stays ship shape. Sometimes both sides will come up with some sort of cooperative fence-fixing arrangement. Sometimes arbitrary decisions made by the people long ago when they first built the fence indicate responsibility. (For your reference, if the wire is on your side of the fence, then by default, it's probably your problem to fix, when in doubt.)

All too often, neither side claims responsibility and nobody even thinks about working on that fence until an obvious hole shows up. And even then, the fix is a quick one, not made to last, just intended to make the worst hole less of a hole and more resemblent of a fence. This usually leads to one or the other, eventually, taking charge and just rebuilding the fence.

But all that being said, the fence brings the neighbors together. After all, the most used part of any fence is the gate. Good fences make good neighbors, and gates can make good friends.

Audra has had some of the best conversations over the fence.

A REASON TO GO SOMEWHERE

Agriculture is not an occupation that takes a lot of time off---and the people that farm and ranch are sometimes unsure of the meaning of the very concept. Getting them to travel is like replacing sickle teeth on a combine. The only surefire way to get them off the place is to give them a good reason.

In case you're wondering, 'good reason' translates to 'work-related' in ag-speak.

So, when the combine breaks down and the nearest part is two-hundred miles away, off they go. No hesitation.

As a kid, I probably enjoyed these incidents of disrepair a little too much. I day-trip to a foreign implement house where they probably had a different selection of ERTL toy-equipment that I was never going to get, but could, at least, admire? I'm in. I bet we'll even stop for cokes. Maybe ice cream if the stars are aligned.

But the biggest trip is to the annual Farm and Ranch Show.

It resembles a vacation in the sense that the whole family would go together. Everything shut down, no one stayed behind to keep working, and all your friends, grandparents, and everyone else you know were mostly doing the same.

To a kid, the farm show is like a theme-park. There are tractors, sprayers, and things that you've never seen before---and you can climb up on them, check out the cabs, and wonder why the ones you spend all your time on at home are less...starship-seeming.

Your folks won't be buying you that amazingly tech-imbued tractor to spend your summer on, but they might pick up a few stocking stuffers. Fencing tools are the perfect thing. Who doesn't want a snazzy stretcher that works so well that you'll never have an excuse to leave a wire sagging again? Or a t-post clip manipulation doohickey that even a toddler can use? If you were really good, you might even get the revered new cow-poker that everyone admires.

The best part, though, is the swag. You don't have to convince anyone, you just have to be willing to carry it all day as you walk miles of farm and ranch equipment and accessories. Yardsticks, candy, and maybe a sweet ball cap, you load yourself down with things you don't need and brochures on things that you both need and want, but can't afford. The farm show will make you wonder who it is that can buy this stuff new? And you can go home and know that in a few years, there will be something better, and you can get this stuff, used.

Audra Brown does judge you on the quality of your complementary yardsticks.

IT'S NOT JUST THE MILES

Modern farming involves a lot of driving.

There's the drive to and from the field, which is likely to vary from no miles to fifty miles, depending on the spread of your operation. I'm not saying there aren't further fields, but I'm going with confidence intervals, not absolutes today.

Then there's the driving in the field. Back and forth and back and forth, or maybe around and around in a circle, or in certain cases, the patented D-Pattern. This is not only a lot of miles, but is done at a rather slow pace. I ran some quick estimations and by those figures, the average non-row-crop operation in the raising of a crop of wheat, on a quarter-section circle…might as well be fifty miles.

Fifty miles. That's not too far, right? An hour of driving time on the highway, sometimes less. But about the fastest operation, 35 feet at a time, is planting, and if you've got the latest double discs, no problems, a smooth field, and you don't stop for anything, that's 6 hours. Disking is slower, call it 6mph, that's 8 hours and change, and likely to need done more than once a year. Harvest is the kicker, and you hope to go really slow.

On a not so great year, you run the combine at about 3mph on average, in whatever pattern you've chosen.

(Sometimes it's faster, but we don't want to remember those years.)

At that speed, 16 hours, or one average workday, is about how much time a field takes. On a significantly better year, you strain the slowness of the hydrostatic drive and cut along at a half a mile per hour. (Ah, the good days.) That's a hundred hours. That's nearly a whole week of work for just one quarter-section field.

It's not much of a stretch to figure a good five-thousand miles per year per farm-family member, just driving around on the place. Once you get old enough to drive to town, that about triples your accumulated distance.

When the ranch is a hundred miles from the farm and the farm is ten miles from the house, and town is twenty miles, and the John Deere house is thirty, it adds up quick.

I'd be willing to bet that a busy farm kid puts in over a 100,000 miles before they are old enough to get that license that lets you drive off the place.

But if you were paying attention to the numbers, it easy to see that it isn't just the miles, it's also the hours.

Audra Brown prefers to measure the days in meals, but it never adds up.

ENJOYING A JOB NEVER DONE

Agriculture is a never-ending business.

It's an ongoing sequence of sometimes-repeating events that rarely gives you a reprieve. And if you find the breaks too long, you can always find a new facet of the business to fill up your free time.

A farmer with too much time in the winter?

Get some wheat calves and you'll stay busy for those months too.

Farming and ranching both, and you're still finding time to sleep?

You must be efficient indeed. Perhaps you can donate some time and worry to a board or committee or some other regularly scheduled meeting.

Even more time to kill?

I'm not sure if I should recommend the hobby of collecting hobbies, but as it is my favorite pastime, I'll just say that it keeps you busy and is rarely boring. Philosophizing about this compulsion to keep busy, leads me to think about the chicken and the egg... Does the ag-life beget hard-workers or does it just find them?

I myself suggest at least an attempt at enjoying those days or hours of satisfying peace after you finish a job. You know you'll have to do it again next year, but forget that for a moment, and just bask in the glow of the headlights as you cool down the tractor after making the last clean-up pass around the edge of the field.

Don't think about the worms that'll probably show up in a couple weeks because you planted with an early rain. Don't contemplate how good or bad the harvest will be next year. Celebrate a little, even if that just means giving yourself and your family one night of not worrying about what needs done tomorrow and what didn't quite get done today.

It occurs to me that a frequently heard wisdom is something about not dwelling in the past. Well, I'm gonna suggest that the farmer not spend all their time dwelling in the future.

There is no finish line.

The goal is to keep going and do it again. So if you don't enjoy the little victories, you won't have much fun at all.

Don't save up your energy for that next time, that bigger win that won't ever come. The goal that really matters is today's job, well done.

Audra Brown's hobby collection is still growing.

THE LONG VACATION

Summer is not a vacation.

Growing up, I didn't go to school, but I had heard of this mysterious thing known as summer break. Despite a couple of other viable interpretations, it was clear that this was intended to denote some sort of reprieve... rather than what sort of plow you'd be using.

Eventually, I began college, my first experience with the semester system of scheduling---and low and behold, I thought it might be my chance to understand that great vacation I'd only heard of...that other thing that was known as the summer.

After one go around the school-year, the hopeful anticipation that my peers shared regarding that central semester, was not reciprocated by me. In fact, I was in many ways the opposite. I enjoyed the freest time when school was in session. The 20-30 hours a week that class-attendance and homework required was a fine vacation.

There were many isolated hours in between class times to be loose. Not enough time to go home and do any work, but a sweet span of minutes that could be dedicated to sipping a glass of iced tea and perusing a good book. Even working on homework in those breaks was quite a pleasant and painless way to spend the time.

Algebra problems may need to be solved, but they require no parts or grease to fix. Your iced-tea glass does not run near the risk of getting knocked over and spilled due to the rough terrain while your turn the tractor and have no extra hands to keep it upright.

In my day, far gone, a paperback book, a palm pilot, an early ipod touch, a printed puzzle, or a my fancy-smancy graphing calculator---got the attention of that free hand that wasn't driving a tractor.

School was a vacation, or the closest I've ever really come. I looked forward to when it was in session and raced back to its slow-paced world after every so-called "summer break" when I was at home.

Summer was no vacation, with weeks of plowing, planting, and cutting wheat to get done. Spring break was a mixed blessing, with less farm work to do, but sound and rehearsal for the Floyd Jamboree took up the slack, keeping me in fun, but still busy. And Thanksgiving break, no matter how short or long or cold, seemed to be reserved for cattle work that my family had conveniently postponed until I could come.

Mind you, I ain't complaining. But when I think of a summer break, it's still sounds like a plow.

Audra Brown writes only fiction when she writes about a real summer vacation.

DON'T FALL OFF

Equipment breaks down, that's just the way it is. A combine, for example, has a lot of moving parts, and it breaks down less often than you'd think for such a complex machine. But when it does, it's usually a bear to work on.

You end up crawling around inside, surrounded by pointy, pokey bits that look like they belong in a medieval torture chamber. You ignore the unbidden images of what would happen if they were to start moving and ignore the discomfort of being awkwardly crammed into whatever pokey little space you're in.

The inside jobs aren't the most comfortable, but at least you can take your time.

Other jobs are not so easily accessible. There's one particular chain, that runs the elevator auger that brings the grain up to the bin, and has a habit of getting loose (it is a chain, that's what they do) and jumping off the sprockets.

To fix it, you've got to thread this long, heavy chain several feet up, until you can grab it at the top through a little twelve-inch hatch on the side. To reach the hatch, you've got to hook a leg or an arm over the side of the bin and hang out over the slick, steep, long-ways-down, side of the combine. It's not big enough or placed such that you can see, but you get your arm inside, get hold of the chain, and then (while ignoring the pain/fatigue of keeping yourself from falling) you've got to blindly thread that big chain around multiple sprockets---in the right pattern.

Eventually, after dropping the chain and starting over a few times, and likely threading the wrong pattern at least once, you get it done and visually inspect that your arms arc both still there, though they probably don't feel like they are in good working order at the moment.

This is actually one of the easier to fix problems that seems to be a regular occurrence on the combine. It usually doesn't take a whole lot of complicated parts, probably a chain-breaker and a master-link, and/or a new wooden idler (it's a fancy chunk of wood with a hole in it. Very hi-tech.) with some wrenches that'll fit the bolt that holds it in.

Hopefully, you aren't broke down long. There's work get back to since you didn't fall off.

Audra Brown didn't fall off and types this with all her fingers.

CRAVINGS IN THE CAB

Cappuccino and a Twinkie.

That might sound like a feast, an odd combination, or just a snack. It depends on the situation.

On the tractor, it's all of the above.

Days spent on a tractor are long. Wise folks will tell you to take your own lunch, probably supper too, lots of snacks, and a full ice chest. But even if you are consistently so prepared, there are always times when somehow, you end up getting hungry.

In a perfect story, every time such tragedy strikes, someone arrives to bring you a brown paper bag, a cold cup of iced tea, and a jalapeno to chew on.

If the clouds are really promising, the blessed delivery-person has also acquired something beyond the basic necessities and surprises you with a candy bar or cookie or coffee.

Just the novelty of something special after days of watching those same plow-bottoms peel back that same dirt, in those same straight rows, with the same ten songs playing on the radio…is sure nice. It can to the point that you almost look forward to cleaning the sweeps off, greasing, fueling, or even…

...no one would admit this, but...sometimes you're even a little relieved when something breaks down.

It's really the little things. A song they play less than once a day, the cold iced tea that came with a hot baloney burrito (beats canned sardines, but...you've got to be good and hungry), a patch of dirt on the floor that you swear looks like the state of Iowa...

The point is, driving a tractor is a long, lonely job, and while it's more comfortable than wrestling a steer, hoeing weeds, or hanging off the high side of some equipment by one arm while trying to blind thread a heavy chain---again---it's monotonous and makes you appreciate the excitement of the more unpleasant, but exciting jobs that come around.

So, remember the tractor driver. They may not be having an exciting day. Listen when they come home and tell you about a big a clod the plow turned over. No, it's not the sort of breaking news you'd see on television, but it was probably the highlight of the tractor driver's day.

And if you really want to make the day, take 'em something different. A snack they don't need. You'll get a hug bigger than on Christmas if you show up in the field with that Twinkie and a Cappuccino.

Audra Brown, personally, would just rather have two coffees and no snacks, thanks.

IN-DEMAND HANDINESS

Some skills are perishable and some skills go out of style, but the level of competence you achieve working as a farmer of rancher, tends to be pretty stable.

It is a hard-won handiness that is always in demand, no matter how far you may stray from the path of agriculture. As we've discussed before, the concept of a vacation is quite foreign to the dedicated farmer or rancher, and indeed, alien to the man or woman who is both.

There are those, who, for one reason or another, leave to pursue the lofty idea of a job that is less dangerous, well-air-conditioned, and not only defines the concept of vacation, but may even pay you to not be there for a few days a year. This dream is usually disappointing in its mundanity compared to the ever-changing challenge of the open field, but sometimes a cushy town job is just what the doctor ordered for a little while.

You do your thing and punch nothing but keys for a while and then the generous two-week, time-off for the holidays is finally here. The plan is to not have a plan, but catch up on the neglected sources of joy that you call hobbies, find and conversate with friends and family for longer than a lunch-hour, and generally enjoy the freedom of having no job that just has to be done for a few days.

This is, of course, a dream.

We all know that vacations do not exist for the agriculturally-involved, and once you've mastered some aspect of the agricultural arts, your skills will forever be in demand---and convention dictates that you can't say no to a pleasant request for expert assistance.

And so, you lay out your paint brushes and start to pour a glass of egg nog, but before you can say, "no," you answer the phone and return to the field to practice your unperished skills.

As if no time has passed, you fall into gathering formation with the other pickups and gather the cattle. The sorting alley is a dangerous, quick, but comfortable place. Your vacation is work, but is it so bad to do what you can do so good?

The worst is the eating, as you've grown accustomed to regularly scheduled meals…

And what do you tell the people who ask that inevitable question about what activities you engaged in in the in-between? "I saw my family for the holidays and now it's back to the long vacation."

Audra Brown is a practiced practitioner of the agricultural arts.

THE SOUNDS TO WHICH YOU CAN AND CAN'T SLEEP

Sounds are funny things. They are always there, but depending on what you are used to hearing, you may or may not notice them.

Personally, I find the sounds of the city to be less than not-noticeable when I'm in a more urban location and trying to sleep.

It's obviously not the decibel level, as the sounds are often faint. But there is something uncomfortable about the sounds of all those people out there, doing all their people things.

On the other hand, I reckon the sound of a corral full of freshly weaned calves, bawling all night a couple hundred feet away---might not be the most sleep-inducing experience for people other than me.

Cattle mooing, corrals creaking, wind blowing, and the occasional snuffing-sneeze of an ornery horse. Those are pretty comforting sounds in the particular spot where I grew up.

When the cattle aren't so close, or so loud, you can fall asleep to the not-so-distant chorus of coyotes howling.

When things are moving faster and you find yourself on

the tractor at apparently odd times of the day and night, you'll probably catch a few winks when you can. You can put your feet up on the dashboard and find somewhere to kinda stash your head, then sleep to the buzzing ambiance of a diesel engine throttled down.

On that note, in addition to the ability to sleep to the sounds of agriculture, it remind me of some of the less and more comfortable places I've been known to find a few minutes of rest. A good nap is more about how much you need it than where you find the time.

Under the trailer, or the pickup, or an empty bag of peanut-seed, sand underneath and hopefully having somehow found some shade. On a sandstone slab, next to a cliff, under a scrubby juniper that at least kinda shaded my head. On the back of horse, in the back of a pickup, in the floor of the combine, in the crook of a tree, or leaning back against a fence post somewhere.

Back on the topic of loud, I'm pretty sure I've stretched out at the sale barn and caught more than a few winks. One of the best places, as long as you don't get hauled off, is up on a loaded trailer of peanuts.

But a hotel? With the lights and sounds of all those people running around? That takes a lot more exhaustion to get any decent sleep.

Audra Brown has napped in some pretty remote places.

WAITING FOR SOMETHING TO GO WRONG

There are a lot of moving parts on a tractor, or a combine, or any of an assortment of important pieces of equipment. There are standard types of preventative maintenance that you can (and better) do to keep the obvious parts from unnecessarily breaking down. But, no matter what you do, it'll eventually stop working. It's what moving parts do.

You grease the pivots, bearings, ball-joints, and anything else that turns, rubs, or looks at you funny. You check and change the various fluids. Engine oil, hydraulic oil, antifreeze, refrigerant, ice-chest... But you can't go through every part and piece each morning before you start to work. So you drive, until it breaks down---or you finish.

There are times when things don't go wrong for a whole job. You start it up, you drive, and then you put it away and go on to the next thing.

That'd be nice.

The state of mind you have to have is a bit of a paradox. Expect the worst, but don't worry about it. Don't dwell on the inevitable breakdown, but don't be surprised or get upset when it does happen.

Of course, I will admit, that while all equipment is gonna break, there are significant differences when it comes to the frequency, probably severity, and how easy it will be to fix.

The average problem requires a part, some tools, and approximately two people who have some idea what they are doing. Many times, this is a common enough occurrence that the part you need is nearby. In that case, the expected downtime is minimal, less than a day, just however long it takes to actually change the part out.

Otherwise…the part may be somewhat farther away and most of your time is spent locating and acquiring said part rather than putting it on the broken equipment. If you're really lucky, a neighbor has one. If you're just a little less lucky, a local dealer has what you need. (By local, I mean anyone within fifty miles.) If you're on the naughty list for some reason, you may be looking at a significant distance between you and what you need..

Often in these cases, you are presented with a choice. Get the part shipped in (to that local dealer) or go get it yourself. Anything within 200 miles is definitely up for consideration if you broke down early in the morning. If you can get there before five, and get back before the next day, you might get running again just a little faster.

Audra Brown has made a few parts runs.

PICKING THE RIGHT PICKUP

We live in a wide land where automobile transportation is not a benefit, but a necessity.

Rules for dealing with cars are almost as important as ways of dealing with people.

And dealing with people's cars and people in cars, is pretty much the epitome of need-to-know.

Getting into someone's vehicle without permission or invitation is a pretty serious line to cross, in general. And I certainly don't recommend ignoring that important standard of conduct in your everyday life.

But, as with many customs, things can be a little different on the farm or out on the ranch.

I'd say you never need an invitation to jump on somebody's tractor to bring them lunch, a snack, or something cold to quench their thirst.

But then again, you might say that stopping to not run over you is an invitation that is clear enough.

The only time this would be real iffy is when the thing that they are doing to the crop is one that must be done at speeds slow enough to safely hop on and off without actually stopping.

If they are running a peanut-thresher down rows that will hold the tractor on course, I'd recommend waiting until they hop off to check around the back, and be sitting there when they return.

Though, be sure to have the snack clearly visible to prevent sudden retaliation in response to such a surprise. And be ready to give them a hand so that they don't jump the wrong way and fall off.

When you're not the one delivering sustenance, then the thing to keep in mind when looking for an automobile to invade is knowing who is likely to have the best supplies and a back seat that is empty enough.

When it's a cold day at the farm auction, the protocol is at it's most relaxed, but also in demand.

Nobody cares who's in the driver's seat of the pickup that is nearest, you only need to look for a head in the seat that you are planning to soon occupy.

Open the door, close it quick to be polite, and if you picked the right rig, you'll find friends, maybe some egg nog that includes the rum, and at least one sharable liquid-holding device.

Audra Brown has so far had good luck and ended up hopping in the best stocked pickup.

CATCHING a CHRISTMAS TREE

Oh, Christmas tree, oh Christmas tree, somewhat flat on one side are you.

I'm a firm proponent of the idea that nothing is impossible. But, while anything can be accomplished, that does not always translate into the job getting done in an optimal manner. Sometimes, the best you can do is not quite as good as you'd hoped for.

And so it was with the Christmas tree hunt one year.

Me and the kid and the one in the middle had been dispatched to the ranch two counties over to feed the cows and check the water so that they ought to be good until Christmas day was over.

Our secondary mission, given to us by our mother, was to also bring back a Christmas tree from hairier side of the place.

The pickup we were in, was the one with the camper-shell, which is nice in the cab with heated-seats and satellite radio to enjoy, but not immediately the best for tree transportation. Hooked on the ball was the feed wagon full of cow cake that we were draggin' all over the place as we made the rounds to find the all the cattle.

Now, the part of the ranch that has trees suitable for

Christmas is a long ways from the entrance and on the later part of the normal feed route. We made it there before dark, still draggin' the wagon, and as difficult as it is to find a decently pointy tree in the daylight, it's much closer to impossible in the dark.

Clearly, the only option was to find a tree now and not later.

We found a decent specimen, nearish to the road, and we went after it with the oscillating saw that into the back we'd throw'd. It works like a charm---when the battery is full, and today it was not, so we had to change tacks and pull.

The chain around the stump, connected to the wagon, did a fine job of making the tree mobile. But the tree was actually quite big when taken as a whole and wouldn't fit in the back and we were not quite enough to get it up the six feet or so onto the back of the wagon. So the several miles back to the nearest powered saw, the tree stayed where it was, hooked onto the back with a chain.

We did get it back, and we did get it sawed. It may have been a little flat on one side, but that just made it more perfect for putting up against the wall.

Audra Brown is just taking an axe next time.

TOOK MY HORSE TO TOWN

I can't say what the county fair looks like to anyone else, but I can share the perspective of a farm kid from a few years ago.

The fair was a wonder and an event, of that there is no doubt. It came around once a year, at a time when hot and monotonous are common modifiers for what I'd been doing for the last few months.

It comes at a time when the farm and the ranch is plenty busy, but the fair was a big enough reason to take a few hours off. (And that alone makes it pretty exciting to a farm kid.)

If you remembered to gather up a some pictures or projects that you'd made the year before, and could get off work for a bit on entry day, you could vie for ribbons if you figured out what category your stuff ought to be in.

I'm not the kid who did 4H or showed animals all day long. Just getting to the fair a couple of times during the week was hard enough to manage. I didn't have show cattle, I just had cattle and heck if I was going to catch one and wash it off.

You got to see other farmers and ranchers and such folks as don't group up in town very much.

Young me didn't understand how my parents could stand and talk to someone for two or three hours when there was all kinds of colorful rarities to walk around and observe.

(I'm not kidding either,
young me timed it more than once.)

But one of my favorite events was the junior rodeo. Once a year, those of us who had working horses, but didn't have time to rodeo regularly, got to play.

The first year, my aunt (who knew her stuff) taught me and my cousins the basics. I can't say I was ready, but at least I knew how it worked.

It's impossible to say who was more nervous and excited when the big night came, me or my horse who hadn't been off the ranch since before I was born.

My horse sure didn't know the patterns by heart, I had to steer him through every turn, but what I lack in polish, I somewhat made up for in guts. I never won my once a year rodeo, but I had fun and was far from last in my favorite event of goat-tying. I haven't rodeoed since they stopped having it, but I'll never forget the feeling, waiting for your turn.

Audra Brown won a bucket and a lead rope and still has both of 'em.

145

THE NERD WHO WORE OVERALLS

There is a thought that people think about growing up on the farm.

They imagine many things, but I suspect that a library isn't the first image that pops into the brain. Some of you may have gotten there quickly and I'm glad you understand.

But, for clarity, let me paint you a picture.

The room was built with one full wall of bookshelves and a few shelves that weren't really optimized for the storage of literature---being too deep and neither tall nor short enough.

Less than half the room was set for books, and yet, it was destined to be defined as the Library.

The works of L'Amour, Asimov, Clavell, Clancy, Cussler, Ludlum, and many more filled the shelves. Stacks hid the neat rows of books that were shelved as intended.

The knowledge and adventure grew and grew, but even with the addition of shelves in pretty much any place that would have them, there was never enough room to hold the collection of bounded paper that contained so much to be experienced.

My love for books was powerful and sudden, and it is still a passion that has dimmed only when it is shaded by the passion that it birthed---the passion to not only read, but to write the stories.

The library at home was where it all started, then the library in town when I could manage to get there. (And I still have a bit of a pain when I remember that there was a limit on the number of books I could take home…) My own collection developed from a shelf, to the walls of my room, to most of my house's walls, to a semi-truck loaded with a number that will be difficult to top.

On the tractor, on a horse, or under some piece of equipment with my back in the sand when it was the only shade to be found...I read a lot.

I read more than most people I knew, with the exception of my dear dad, so it wasn't a thing that I got to relate with to others or go places to talk about.

But I still remember how much fun it was when my dad started taking me to the Williamson Lectures. Now Jack is gone, but the fun still remains. I'll be there tomorrow because it is a special day.

Audra Brown is a sand-bookworm

WHERE EVERYBODY KNOWS YOUR PICKUP

People are so complex as to be impossible to completely describe, but we need to differentiate them. We use names, faces, and the way that they walk, talk, and dress to keep track of which is who and so forth. These features are all most telling only when you are in close contact with the human you wish to identify.

But when the person-density is low, as it is on the plains, and distance is significant such that walking is not much used as a common form of transportation.

You don't pass by people on the side-walk, you pass their vehicles on the road.

Therefore, your personal mode of transportation is more notable and unique than your face, quite possibly.

There are the obvious differentiations---make, model, color, and the sound of the exhaust. But those are not enough to tell apart the many white, 4-door, Fords that were what was on the lot when the old pickup finally broke down for good last time. No to mention the farm-common single-cab with a flatbed and a cake-feeder.

What makes each pickup the unique reflection and four-wheeled face of the owner? The dents, of course, and the stuff in the back.

You learn to see what welder they have and the color and configuration of their toolbox. You know that so and so has his hose-reel on the left and that good ol' boy has his mounted on the right.

There is the type and condition of the grill-guard, and the shape one's bumpers tend towards as they encounter obstacles and things to pull. Some are always more prone that others to look like frowning chrome.

Sometimes, you get it easy and can tell by the cow-dog.

But if not, and you still haven't quite narrowed it down, then we must consider the less intentionally acquired features like dirt and dust and spit. The color of the dirt, and how long it collects; the dents and the scratches and the two bumper stickers that may or may not still be relevant to current events.

You can tell a lot by the pickup that someone drives, and when that's how you see people, you can see it impressively fast. All those bits that we just laid out are often passed at significant speed.

An ag-person's pickup is a lot like his face, worn by the weather and shaped by the needs of the land. But at least with a pickup, you do occasionally get to trade up.

Audra Brown will wave from her pickup seat if you meet her on the road today.

THE INVESTMENT

Farming is an investment, not only of capitol and time,
but of the soul.

It is not an enterprise suited to dipping a toe in to check
the water. It cannot be done halfway or in part. Most
businesses can start in a place so small and a scale
so reduced that the risk is almost irrelevant and the
investment is more time than anything else. You can
start building computers in your garage and gradually
grow so big that I don't even need to use any proper
nouns and folks will still have an idea who I'm talking
about.

You can't start a farm in your backyard.

There's not enough room to park the tractor.

There is a minimum set of equipment needed to begin
any farming adventure. The same set, at the same cost,
is the minimum required to farm ten acres or a ten-
thousand. Alas, unless you have a million or so laying
around to get that starter set of equipment debt-free,
you'll need to pay for it by using it. Ten acres on the high
desert ain't gonna cut it. They just don't make mortgage
terms that long, and the human lifespan isn't that
impressive. So, go big or maybe just go back to doing
something a little less extreme. Farming isn't for the faint
of heart, or anyone in any sort of hurry.

But, if you've got the soul of farmer, then you'll dive in and good luck to ya, my friend. It's a beautiful life, making your own 80-hour-a-week schedule. (Five 16 hour days? Seven 12 hour days? Or maybe just a couple of all-day-and-nighters?) No boss telling you what to do, just the one way upstairs keeping you on your toes by playing with the weather.

On a good year, the rains come at the right times, the bugs don't get out of hand, and commodity prices aren't lower than they were forty years ago. Pay off the operating loan, fix all the equipment, fill the diesel tank, and maybe sneak off for a few days of vacation.

The rest of the time, you can scrape by and get close enough to breaking even to call it breaking even and hang on for another go-round.

To be honest, without a love of the field and a manic faith that it'll get better (or not worse), the straight numbers are hard to line up.

For anyone gung-ho enough to make it farming, there were a lot of easier ways to make a living. But for those who love it, they wouldn't change a thing and they'll never be able to go back.

Audra Brown is possibly too fond of A/C to be perfectly faithful.

ABOUT THE AUTHOR

Audra Brown is more than a little awkward when writing a introduction to herself, which may be considered odd due to the fact that she considers herself very handy at most things--including writing.

She writes a lot, mostly fiction in the areas of not-romance. She's had several stories published, and has even been paid for one particular horror story. She is in the middle of the slow process of publishing her first novel and getting this book ready to go.

In the meantime, she writes more columns, novels, stories, and screenplays.

Writing is a lot like farming, lots of hard work and long hours up front, and then a long wait and a lot of praying, and then maybe eventually (or maybe not) the farmer gets paid. Farming actually seems easier sometimes.

Maybe that's why she still does it.

Though she'd consider her business to be more cattle-oriented, Audra is, and always will be, a farm kid.

She was driving dump buggies about the same time her contemporaries were starting school. She did not join them, instead spending her school years starting her own cattle operation and working on her family's farming and ranching operation.

Her education was unorthodox, and is still largely undocumented, but in no way lacking. Mathematics, science, and reading went hand in hand with diesel mechanicing, part fabrication, and all the other exciting problems that plague a farm.

Somewhere along the way, she made room for a few hobbies. Reading, leatherwork, competitive shooting, and muscle car collection and restoration--to name but a few. And one more that is perhaps one of her crowning achievements so far.

Martial Arts, specifically, Taekwondo, is a passion she has followed almost as long as she has known how to drive a tractor. Presently, she is a 5th degree black belt, 5-time member of Team USA, and repeating World Champion Power Breaker.

She's been around the turnrow, and around the world, and hopes to be around a few more times at least.

The important thing about Audra today, is that she is the author of this collection of reality-inspired anecdotes and hopes that you all enjoy the exploits and opinions that she will be sharing in this and future editions. Howdy ya'll!

Ready for more adventure?

Navigate to

2ePublishing.com

to stay up to date on all the new books including past and future *Adventures in Agriculture.*

For more from Audra Brown:

www.audra-brown.com
www.al-brown.com

Instagram: @caffeinated_polymath
Twitter: @thegreenninja24

Also stay in touch by following her at:
www.facebook.com/ToughTarget

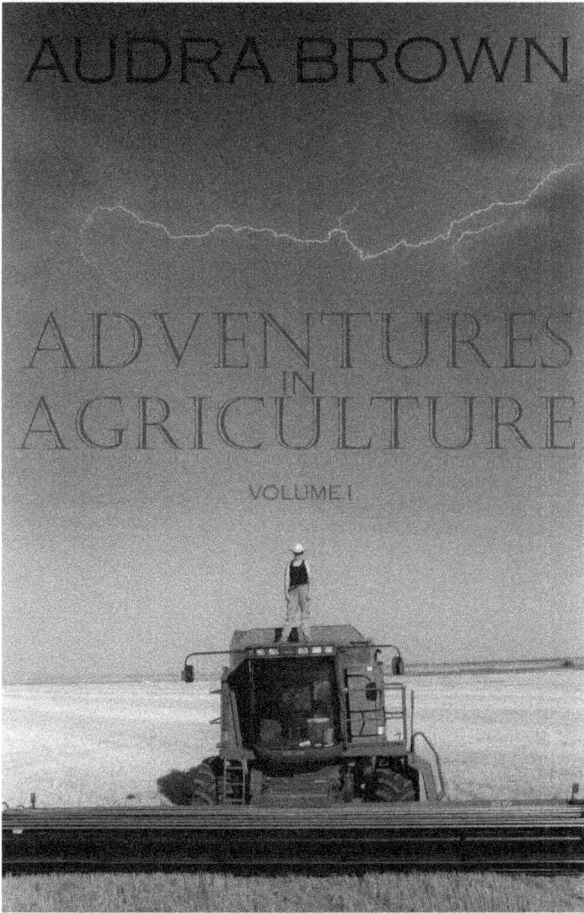

WHAT READERS SAY:

"Quite a hoot!"

"This is a well-written, fun, and entertaining book!"

"Audra Brown is like the Baxter Black of farmers."

TOUGH TARGET

WHAT READERS SAY:

"...it's like a western, spy, adventure novel!"

"The layers and twists of the story made this book very hard to put down."

"The author takes a compelling hero and places her in a land where survival is a daily job."

"Josey Jackson's character will strap you into her passenger seat and take you for an unbelievable ride!"

"People always say the books they like are romps, but it'd be more accurate to describe Brown's 429 to Yuma as a joyride. A delightfully quippy protagonist and wry description make this fast paced read come alive--readers might as well be flying down New Mexico highways right alongside Josey Jackson, wishing they'd brought a fire extinguisher and some burritos of their own. 5 Stars"

www.ingramcontent.com/pod-product-compliance
Lightning Source LLC
Chambersburg PA
CBHW021403090426
42742CB00009B/980